FOURIER TECHNIQUES AND APPLICATIONS

FOURIER TECHNIQUES AND APPLICATIONS

Edited by

John F. Price

School of Mathematics
University of New South Wales
Kensington, New South Wales, Australia

Plenum Press • New York and London

Library of Congress Cataloging in Publication Data

Main entry under title:

Fourier techniques and applications.

"Based on the Short Course on Applied Fourier Analysis, held August 29–30, 1983, and on the Conference on Fourier Techniques and Applications, held August 31–September 2, 1983, in Kensington, New South Wales, Australia"—T.p. verso.
Includes bibliographical references and index.
1. Fourier analysis—Congresses. I. Price, John F. (John Frederick), 1943- . II. Short Course on Applied Fourier Analysis (1983: Kensington, N.S.W.) III. Conference on Fourier Techniques and Applications (1983: Kensington, N.S.W.)
QA403.5.F68 1985   515′.2433   85-19255

ISBN-13: 978-1-4612-9525-9   e-ISBN-13:978-1-4613-2525-3
DOI: 10.1007/978-1-4613-2525-3

Based on the Short Course on Applied Fourier Analysis, held August 29–30, 1983, and on the Conference on Fourier Techniques and Applications, held August 31–September 2, 1983, at the University of New South Wales, Kensington, New South Wales, Australia

©1985 Plenum Press, New York
Softcover reprint of the hardcover 1st edition 1985
A Division of Plenum Publishing Corporation
233 Spring Street, New York, N.Y. 10013

PREFACE

The first systematic methods of Fourier analysis date
from the early eighteenth century with the work of Joseph
Fourier on the problem of the flow of heat. (A brief history
is contained in the first paper.) Given the initial temperature at all points of a region, the problem was to determine
the changes in the temperature distribution over time.
Understanding and predicting these changes was important in
such areas as the handling of metals and the determination of
geological and atmospheric temperatures.

Briefly, Fourier noticed that the solution of the heat
diffusion problem was simple if the initial temperature distribution was sinusoidal. He then asserted that any distribution can be decomposed into a sum of sinusoids, these being
the harmonics of the original function. This meant that the
general solution could now be obtained by summing the solutions of the component sinusoidal problems.

This remarkable ability of the series of sinusoids to
describe all "reasonable" functions, the sine qua non of
Fourier analysis and synthesis, has led to the routine use of
the methods originating with Fourier in a great diversity of
areas – astrophysics, computing, economics, electrical
engineering, geophysics, information theory, medical
engineering, optics, petroleum and mineral exploration, quantum physics and spectroscopy, to name a few.

With this in mind it was decided to have a week of
applied Fourier analysis at the University of New South Wales
with the idea of bringing together a wide cross-section of
users of Fourier and spectral methods and mathematicians
interested in helping to develop and refine new methods. On
the first two days was a short course "Applied Fourier
Analysis" intended to give people an introduction to some of
the basic tools in the area. It was attended by approximately 100 people.

On the last three days was a conference "Fourier Techniques and Applications" in which twenty-five papers were presented. These were in the areas of filtering (for example, recursive digital filters, duly ratio filters), signal analysis (sampling theorems, uncertainty principles, speech analysis, surveillance systems), image processing (phase restoration, holographic interferometry), geophysics (petroleum exploration, potential fields, cepstral techniques), spectral estimation (maximum entropy methods, engineering applications, sonar analysis), short-term spectra (traffic signal coordination) and numerical methods (partial differential equations). Also considerable care was given in the selection of seven invited speakers. Their talks provided overviews of developments in their areas and served as a framework for most of the contributed papers.

In response to the many inquiries for copies of the lectures and talks it was decided to produce this volume. Part I consists of the lecture notes for the two-day course and Part II five of the invited papers and abstracts of two. In most cases substantial bibliographies have been provided so that the interested reader may first gain a general introduction to any of the areas covered and then follow up, via the references, with a more detailed study.

It gives me pleasure to thank the Committee (Gavin Brown, John Fenton, Harvey Holmes, Christopher Phillips and Neville Rees) for all the work they put into ensuring the success of the week. Financial support was received from the Australian Mathematical Society, the Radio Research Board, the Sydney County Council and the School of Mathematics, University of New South Wales. My thanks for the typing to to Mayda Shahinian; her perseverance overcame many obstacles. Finally I thank all the speakers (particularly those who contributed to this volume) and all the participants.

CONTENTS

THE LIFE OF JOSEPH FOURIER

R.N. Bracewell

Department of Electrical Engineering
Stanford University
Stanford, California, U.S.A.

Baron Jean-Baptiste-Joseph Fourier (21 March 1768 - 16 May 1830), born in poor circumstances in Auxerre, introduced the idea that an arbitrary function, even one defined by different analytic expressions in adjacent segments of its range (such as a staircase waveform) could nevertheless be represented by a single analytic expression. This idea encountered resistance at the time but has proved to be central to many later developments in mathematics, science, and engineering. It is at the heart of the electrical engineering curriculum today. Fourier came upon his idea in connection with the problem of the flow of heat in solid bodies, including the earth.

The formula

$$x/2 = \sin x - \frac{1}{2}\sin 2x + \frac{1}{3}\sin 3x + \ldots$$

was published by Leonhard Euler (1707-1783) before Fourier's work began, so you might like to ponder the question why Euler did not receive the credit for Fourier's series.

Fourier was obsessed with heat, keeping his rooms uncomfortably hot for visitors, while also wearing a heavy coat himself. Some traced this eccentricity back to his three years in Egypt where he went in 1798 with the 165 savants on Napoleon's expedition to civilize the country. Prior to the expedition Fourier was a simple professor of mathematics but he now assumed administrative duties as secretary of the

Institut d'Égypte, a scientific body that met in the harem of
the palace of the Beys. Fourier worked on the theory of
equations at this time but his competence at administration
led to political and diplomatic assignments that he also
discharged with success. It should be recalled that the
ambitious studies in geography, archaeology, medicine, agri-
culture, natural history and so on were being carried out at
a time when Napoleon was fighting Syrians in Palestine,
repelling Turkish invasions, hunting Murad Bey the elusive
Mameluke chief, and all this without support of his fleet,
which had been obliterated by Nelson at the Battle of the
Nile immediately after the disembarkation. Shortly before
the military capitulation in 1801, the French scientists put
to sea but were promptly captured with all their records by
Sidney Smith, commander of the British fleet. However, in
accordance with the gentlemanly spirit of those days, Smith
put the men ashore, retained the documents and collections
for safekeeping and ultimately delivered the material to
Paris in person, except for the Rosetta stone, key to Egyp-
tian hieroglyphics, which stands today in the British Museum
memorializing both Napoleon's launching of Egyptology and his
military failure.

     Fourier was appointed as Prefect of Isere by Napoleon in
1802 after a brief return to his former position as Professor
of Anlysis at the Ecole Polytechnique in Paris. His duties in
Grenoble included taxation, military recruiting, enforcing
laws and carrying out instructions from Paris and writing
reports. He soothed the wounds remaining from the Revolution
of 1789, drained 80,000 square kilometers of malarial swamps
and built the French section of the road to Torino.

     By 1807, despite official duties Fourier had written
down his theory of heat conduction, which depended on the
essential idea of analyzing the temperature distribution into
spatially sinusoidal components, but doubts expressed by
Laplace and Lagrange hindered publication. Criticisms were
also made by Biot and Poisson. Even so, the Institut set the
propagation of heat in solid bodies as the topic for the
prize in mathematics for 1811, and the prize was granted to
Fourier but with a citation mentioning lack of generality and
rigor. The fact that publication was then further delayed
until 1815 can be seen as an indication of the deep uneasi-
ness about Fourier analysis that was felt by the great
mathematicians of the day.

It is true that the one-dimensional distribution of heat
in a straight bar would require a Fourier integral for its
correct expression. Fourier avoided this complication by
considering heat flow in a ring, that is a bar that has been
bent into a circle. In this way, the temperature distribution is forced to be spatially periodic. There is essentially no loss of generality because the circumference of the
ring can be supposed larger than the greatest distance that
could be of physical interest on a straight bar conducting
heat. This idea of Fourier remains familiar as one of the
textbook methods of approaching the Fourier integral as a
limit, starting from a Fourier series representation.

Fourier was placed in a tricky position in 1814, when
Napoleon abdicated and set out for Elba with every likelihood
of passing southward through Grenoble, on what has come to be
known today as the Route Napoléon. To greet his old master
would jeopardize his standing with the new king, Louis XVIII,
who in any case might not look favourably on old associates
and appointees of the departing emperor. Fourier influenced
the choice of a changed route and kept his job. But the next
year Napoleon reappeared in France, this time marching north
through Grenoble where he fired Fourier, who had made himself
scarce. Nevertheless three days later Fourier was appointed
Prefect of the Rhône at Lyons, thus surviving two changes of
regime. Of course, only 100 days elapsed before the King was
back in control and Napoleon was on his way to the south
Atlantic never to return. Fourier's days in provincial
government then ended and he moved to Paris to enter a life
of science and scientific administration, being elected to
the Académie des Sciences in 1817, to the position of per-
manent secretary in 1823, and to the Academie Française in
1826. He never married.

At the beginning mention was made of Euler's formula.
The formula is correct for $-\pi < x < \pi$ but not for other
ranges of x. The right side is the Fourier series for the
sawtooth periodic function f defined by

$$f(x) = \begin{cases} x/2 & \text{for } -\pi < x < \pi \\ 0 & \text{for } x = \pi \\ f(x+2\pi) & \text{for all } x. \end{cases}$$

Fourier wrote, around 1808-9, "the equation is no longer true

when the value of x is between π and 2π. However, the
second side of the equation is still a convergent series but
the sum is not equal to x/2. Euler, who knew this equation,
gave it without comment." (From J. Herivel, "Joseph
Fourier, the Man and the Physicist", Clarendon Press, Oxford,
1975, p.319.)

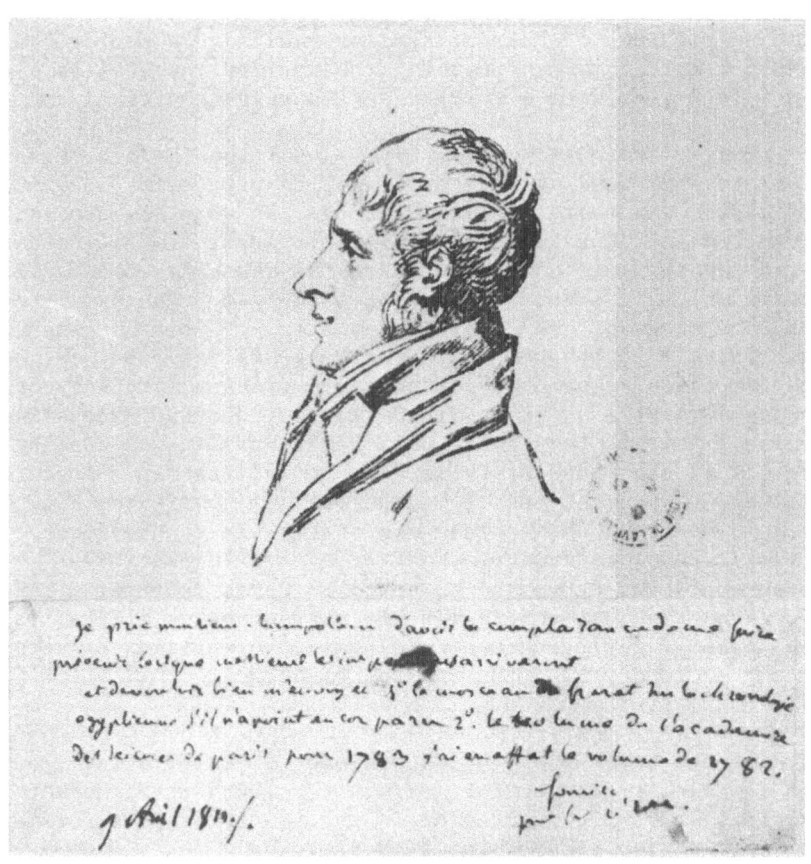

Sketch of Fourier as a young man by his friend
Claude Gautherot, an artist and sculptor. It is taken from
a print held in the Municipal Library of Grenoble, France.

PART 1

APPLIED FOURIER ANALYSIS

LINEAR SYSTEMS, FILTERS AND CONVOLUTION THEOREMS

J.W. Sanders

School of Computing Sciences
New South Wales Institute of Technology
Sydney, N.S.W., Australia

0. INTRODUCTION

The purpose of this paper is to introduce the Fourier transform from the viewpoint of System Theory, to motivate some of the basic techniques associated with it, and to give the Fourier transform´s fundamental properties.

Of the many introductory texts which use a similar approach, the reader may care to consult Kailath [5] and Papoulis [8]; for more information on the Fourier transform (s)he may like to consult Bracewell [1], Dym and McKean [2], Feller [4], Volume 2 and Papoulis [7]; and for a slightly less applied introduction, Edwards [3].

1. EXAMPLE

We begin with a straightforward problem and its solution. One of the basic tools in computing is the (pushdown) stack: given an input queue 1,2,...,n you may <u>push</u> the digits onto the stack in numerical order and <u>pop</u> the most recent digit off the stack at any time, onto the output queue (a stack is often called a last-in-first-out device). For instance in the diminutive case n = 3, the rearrangement 231 may be obtained as follows:

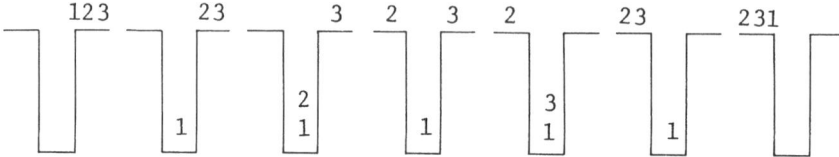

In terms of pushes and pops, we can record the action above:

push, push, pop, push, pop, pop.

The problem we wish to solve is: exactly how many rearrange-
ments of the input queue 1,2,...,n are possible using a
stack? Certainly not all n! permutations occur: for
instance in our diminutive case above, 312 cannot appear on
the output queue (why?); yet the other 5 permutations of
1,2,3 do occur.

At first sight it´s hard to enumerate the required
number explicitly, so we postpone the effort and give the
required number a name – a_n. Having taken this step we can
do something more subtle than calculating a_n directly: we
can get a recursive equation for it. Indeed the stack permutations of 1,2,...,n are obtained by first stack-permuting
1,2,...,j and then stack-permuting j+1, j+2, ..., n. Of
the first there are a_{j-1} permutations (since the digit 1
plays the part of bookending the other j-1 digits) and of
the second there are a_{n-j}. So

$$a_0 = a_1 = 1$$

(1)

if n > 1, $$a_n = \Sigma_{j=1}^{n} a_{j-1} a_{n-j} = \Sigma_{j=0}^{n-1} a_j a_{n-j-1}.$$

We recognize this as a <u>convolution</u>: given sequences $(b_n)_{n \geqslant 0}$
and $(c_n)_{n \geqslant 0}$ their convolution is defined by

$$(b * c)_n = \Sigma_{j=0}^{n} b_j c_{n-j}.$$

So $a_n = (a * a)_{n-1}$, for n > 1. The time-honoured way to
make use of this fact is to "transform" the sequences to
their generating functions: if

$$B(s) = \Sigma_{n \geqslant 0} \, b_n \, s^n \quad \text{and} \quad C(s) = \Sigma_{n \geqslant 0} \, c_n \, s^n \quad \text{then}$$

$$B(s) \, C(s) = \Sigma_{n \geqslant 0} \, (b * c)_n \, s^n.$$

So letting $A(s) = \Sigma_{n \geqslant 0} \, a_n \, s^n$ we equate coefficients using (1) to deduce

$$A(s) - 1 = s \, A(s) \, A(s)$$

or $$A^2(s) - s^{-1}A(s) + s^{-1} = 0$$

so $$A(s) = (1 \pm (1-4s)^{1/2})/2s.$$

We now have to "invert" this equation to find the generating function form of the right-hand side to reclaim the a_n's.

By the binomial theorem, i.e.

$$(1 + x)^{1/2} = \Sigma_{n \geqslant 0} \, \binom{\frac{1}{2}}{n} \, x^n$$

we see

$$(1 \pm (1 - 4s)^{1/2})/2s = (1 \pm 1 \mp 4s \pm \ldots)/2s.$$

For small s, where the generating function converges, the first of these is unbounded whilst the second is bounded, hence we take

$$A(s) = (1 - \Sigma_{n \geqslant 0} \, \binom{\frac{1}{2}}{n}(-4s)^n)/2s$$

$$= \Sigma_{n \geqslant 1} \, (-1)^{n-1} \binom{\frac{1}{2}}{n} 2^{2n-1} \, s^{n-1}$$

$$= \Sigma_{n \geqslant 1} \, \binom{2n-2}{n-1} \, 2^{-2n+1} n^{-1} \, 2^{2n-1} \, s^{n-1}.$$

So at last we can reveal, by uniqueness of the Taylor series for A, that

$$a_n = \binom{2(n+1)-2}{n+1-1}/(n+1) = \binom{2n}{n}/(n+1).$$

As a check, $a_3 = 6!/(3!4!) = 5$. (Observe that asymptotically, a randomly chosen permutation is a stack permutation with probability $4^n.n^{-1}/n^n \sim 0$.)

In summary we had an unknown formula which converted n to a_n:

and to find the formula, involved convolution, transform and inversion. A black box like this is called a **system** (rather trivial here since both input and output are numbers rather than functions) and we have seen that even for simple systems, these basic tools of convolution, transform and inversion are useful. In fact this system is an old one. It arises in the enumeration of binary trees having n nonterminal nodes (in our context, the trees are the parse trees of the grammar

$$S \to (\text{push } S | \text{pop } S | \varepsilon)$$

— see Knuth, [6], p. 389; in counting the number of ways a polygon can be triangulated — see Feller, [4], Volume 1, p. 283; and in enumerating the random walks in the plane not passing below the x-axis (in our context ↗ stands for push and ↘ for pop)—again see Feller, [4], Volume 1, p.73.

2. DISCRETE SYSTEMS

It really is surprising that the toy system in section 1 exhibited the characteristics of more complex systems. In this and the next section we consider more complicated systems and observe how convolution, transform and inversion all play a vital role.

Suppose we sample a signal (a function of a real variable) at discrete time intervals ..., −2, −1, 0, 1, 2, ... and allow some black box (for instance your CD player) to operate on these digitised signal values to give an output digitised response:

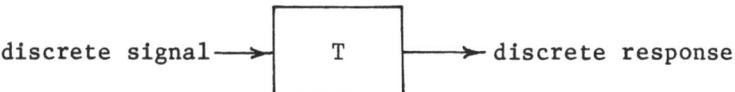

discrete signal——▶ [T] ——▶ discrete response

We call this a <u>discrete system</u>. Without knowing the contents
of the black box we may entertain the following properties.

1. If any signal is made louder, the response becomes
    louder by the same amount: $T(cf) = c(Tf)$.

2. If two signals are superposed, the response equals the
    superposition of the individual responses:
    $T(f + g) = Tf + Tg$.

3. In calculating the response at time N, only values of
    the signal at time $n < N$ are used.

4. If any signal is delayed then its response is delayed by
    the same amount: $T(t_N f) = t_N(Tf)$, where t_N is the
    delay system

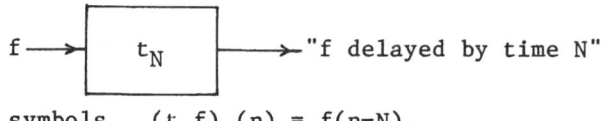

f ——▶ [t_N] ——▶ "f delayed by time N"

or, in symbols, $(t_N f)(n) = f(n-N)$.

5. If two signals differ only slightly then their responses
    differ only slightly: if the signals f_i approach f
    (in terms of say energy or maximum amplitude) then their
    responses Tf_i approach Tf. In the presence of pro-
    perties (1) and (2), this is equivalent to: for some
    constant K, a response is no stronger than K times
    the strength of the signal.

    It is an important step in the modelling procedure to
decide the truth or falsity of these properties for the system under consideration, and often when one property fails,
the system is still approximable by systems satisfying them
all (see for instance section 5). Properties (1) and (2)
together specify <u>linearity</u>; (3) is <u>causality</u>; (4) is <u>time
invariance</u>; and (5) is <u>continuity</u> (beware: many books use
this for what we shall call analogue).

    How can we calculate the effect of a discrete system T?
First we analyse discrete signals. The simplest is a unit
impulse at time n = 0 - let's call it δ:

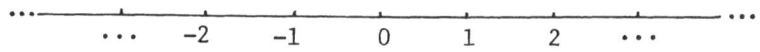

so that $\delta(n)$ is 1 if $n = 0$ and is 0 otherwise.

If f is any discrete signal then we can write it as an infinite linear combination of delays (both forwards and backwards) of δ:

$$f = \Sigma_k \ f(k) \ t_k \delta.$$

So the system's response to the signal f is

$$Tf = T(\Sigma_k \ f(k) t_k \delta)$$
$$= \Sigma_k \ f(k)T(t_k\delta) \quad \text{if} \quad T \quad \text{is continuous and linear}$$
$$= \Sigma_k \ f(k)t_k(T\delta) \quad \text{if} \quad T \quad \text{is time invariant.}$$

Thus we only need to know how the system responds to the unit impulse δ in order to calculate its response to any signal. $T\delta$ is the <u>impulse response</u> and usually written as h. Then we have shown that

$$Tf = \Sigma_k f(k)t_k h = f * h$$

with the definition of convolution given in section 1 (note that δ is an identity for convolution: $\delta * f = f * \delta = f$). In other words: <u>A continuous, time-invariant, linear system acts on a signal f by convolving f with the impulse response h.</u>

Warning: the convolution sum may fail to converge, so this result is only helpful when the input signals f satisfy, for all n,

$$\left| \Sigma_k \ f(k)h(n-k) \right| < \infty.$$

This is often expressed, when the input signals form a Banach space, by saying that h lies in the dual Banach space. Indeed this is a way to give a formal definition of "system".

In terms of the impulse response h, T is causal if and only if $(Tf)(n) = \Sigma_{k<n} f(k)h(n-k)$ which holds if and only if $h(n-k) = 0$ for $k > n$, i.e. $h(m) = 0$ for $m < 0$.

In section 1 we have already seen the importance of broadening our outlook from a discrete signal $(a_n)_n$ to its generating function (or "transform"). This prompts us to put the signal x^n into a discrete system:

$$x^n \longrightarrow \boxed{} \longrightarrow Tx^n = \Sigma_k \, x^{n-k} \, h(k) = x^n \, \Sigma_k \, x^{-k} \, h(k).$$

The term $\Sigma_k \, x^{-k} \, h(k)$ is the <u>system function</u> and is written $H(x)$, so that $T(x^n) = x^n \, H(x)$. Then the system's response to $\Sigma_n \, a_n x^n$ is $H(x) \Sigma_n \, a_n x^n$. Consequently if two systems are put in series then the resulting system

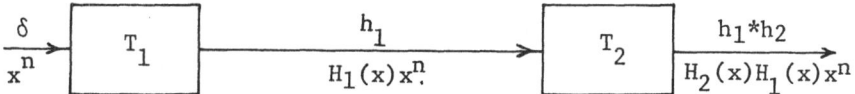

written $T = T_2 \cdot T_1$, has impulse response $h_1 * h_2$ and system function $H_1 H_2$. The system function $H(x)$ is often called the <u>transfer function</u>.

The transform $\Sigma_k \, x^{-k} \, h(x)$ (for complex x) is the <u>Z-transform</u> of h: thus the system function is the Z-transform of the impulse response. In section 1 we transformed a sequence $(a_n)_n$ to get its generating function (which by comparison had a real variable near the origin with exponent +k, and all negative coefficients equal to 0) explicitly, then inverted this to recover the a_n's. We ensured they were unique by checking that $A(s)$ was determined by its Taylor series near the origin. For a discrete system things are similar: given a system function $Z(x)$ (with x complex) there may be several functions h whose Z-transform is Z. However, there is only one having a fixed domain of convergence in the complex plane. How is h to be calculated from Z? We find it neatest to postpone this to the next section, where it will appear as a consequence of inversion for analogue systems.

3. ANALOGUE SYSTEMS

We wish our concept of system to be general enough to convert an analogue signal into an analogue response. Such systems are called <u>analogue systems</u> (or "continuous" systems!):

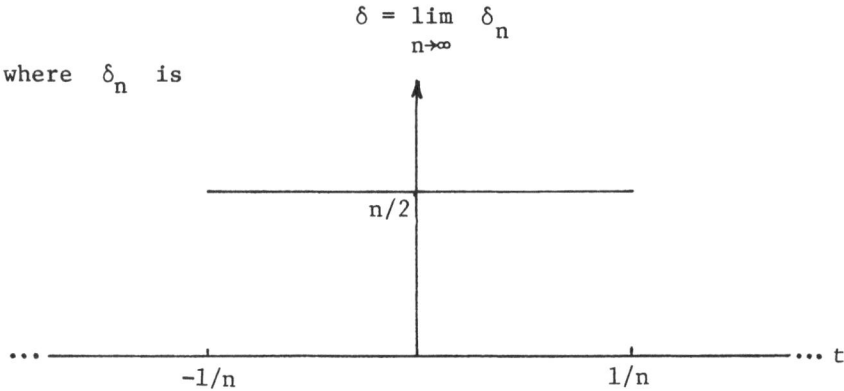

The discrete ideas of the previous section will reapply
provided that we can make them analogue. Firstly the unit
response δ at time t = 0 is now the limit (in a precise
mathematical sense) of a sequence of analogue functions

$$\delta = \lim_{n \to \infty} \delta_n$$

where δ_n is

(infinitely many choices are possible here for the sequence
$(\delta_n)_n$ and continuously differentiable functions are sometimes preferable).

In the discrete case the unit impulse was an identity for
convolution, so that in particular,

$$\Sigma_n \; f(n) \; \delta(n) = f(0).$$

In the present case δ has the analogous (!) property:
δ * f = f * δ = f   so in particular,

$$\int f(t) \; \delta(t) \; dt = f(0).$$

The system's response to δ, the _impulse_ _response_ is

$$h(t) = \lim_{n \to \infty} T\delta_n$$

which can be shown to exist and be independent of the choice
for $(\delta_n)_n$ (see for instance Papoulis [8], p. 96). Again
the convolution property holds (proof by approximating the
integral)

$$(Tf)(t) = \int f(t-u) \; h(u) \; du = (f * h)(t).$$

Yet again, the _system_ _function_ is given by considering the
response to x^t, where now, since we are concerned with
periodic (or almost periodic) signals, we take $x = e^{2\pi i w}$ to

give

$$T\ e^{2\pi i w t} = \int e^{2\pi i w(t-s)}\ h(s)ds$$

$$= e^{2\pi i w t} \int h(s)\ e^{-2\pi i w s}ds.$$

So we set

$$H(w) = \int h(s)e^{-2\pi i w s}\ ds = \hat{h}(w)$$

and call H or $\hat{h}$ the <u>Fourier</u> <u>transform</u> of h. As before, the main feature is the convolution property: if two systems T_1 and T_2 are connected in series then for the resulting system T,

$$T = T_2 \circ T_1, \qquad H = h_1 * h_2, \qquad H = H_1 H_2.$$

For inversion, we are given $Z(w)$ and seek h such that

$$Z(w) = \int h(s)e^{-2\pi i w s}ds.$$

If we observe that $\delta(t) = \int e^{2\pi i t w}\ dw$, (the details of which are best established when h is formally defined but for a check, transform both sides), then

$$\delta = \int e^{2\pi i t w}\ dw \longrightarrow \boxed{} \longrightarrow h$$

where the response is

$$h(t) = (h * \delta)(t) = (h * \int e^{2\pi i t w}\ dw)(t)$$

$$= \int h(s) \int e^{2\pi i (t-s)w}\ dw\ ds$$

$$= \int e^{2\pi i t w} \int h(s)e^{-2\pi i s w}\ ds\ dw$$

$$= \int e^{2\pi i t w}\ H(w)dw$$

and this is the <u>inversion</u> <u>formula</u> expressing h as the inverse Fourier transform of H. As a consequence we get discrete inversion: given $Z(w)$,

$$h(n) = \int_0^1 Z(w) \, e^{2\pi i n w} \, dw.$$

Our claim that the fundamental techniques of system theory can be motivated by section 1 may be wearing a little thin, so it's time for an example. See also section 5.

Example.   An analogue signal can be decomposed into its periodic components $\exp(2\pi i w t)$ and the Fourier transform of the signal at w is its projection onto $\exp(2\pi i w t)$. Then the ideal low-pass filter is represented by a system whose system function is, for some frequency t_0,

$$H(w) = \begin{cases} e^{-2\pi i w t_0} & \text{if } |w| < c \\ 0 & \text{if } |w| > c \end{cases}.$$

To determine the impulse response we merely use the inversion formula

$$h(t) = \int_{-c}^{c} e^{2\pi i t w} \, e^{-2\pi i w t_0} \, dw$$

$$= \frac{e^{2\pi i w (t-t_0)}}{2\pi i (t-t_0)} \Bigg]_{-c}^{c}$$

$$= \frac{\sin(2\pi c(t-t_0))}{\pi(t-t_0)}.$$

4.   PROPERTIES OF THE FOURIER TRANSFORM

The Fourier transform has thus been seen to be of fundamental importance firstly in calculating the system function, and secondly in finding the components of a periodic signal. We now list its main properties, being content to refer the reader to the books mentioned in the introduction for proofs.

I   __Linearity__ $(af + fg)\hat{} = a\hat{f} + f\hat{g}$ for constant a,b.

II   __Translation__ $(t_a f)\hat{}(w) = e^{-2\pi i a w} . \hat{f}(w)$ for any
constant a, where t is
the delay function
$(t_a f)(s) = f(s-a)$.

III   __Dilation__ $(\sigma_a f)\hat{}(w) = |a|^{-1} \hat{f}(w/a)$ for any nonzero
constant a, where σ is the
scaling system
$(\sigma_a f)(t) = f(at)$.

IV   __Conjugation__ $(\bar{f})\hat{}(w) = (\hat{f}(-w))\bar{}$.

V   __Convolution__ $(f * g)\hat{} = \hat{f} . \hat{g}$. (See also X(c).)

VI   __Modulation__ $(e^{2\pi i \gamma t} f(t))\hat{}(w) = \hat{f}(\gamma - w)$. (cf. II)

VII   __Differentiation__

$$(D^n f)\hat{}(w) = (iw)^n \hat{f}(w) \quad \text{and}$$

$$D^n(\hat{f}) = ((-it)^n f(t))\hat{}.$$

If $g(t) = \int_{-\infty}^t f(x)dx$ then $\hat{g}(w) = (iw)^{-1}\hat{f}(w)$ if $w \neq 0$.

VIII __Inversion__ $f(t) = \int \hat{f}(w) e^{2\pi i w t}dw$ if and only if

$\hat{f}(w) = \int f(t)e^{-2\pi i w t}dt$, provided that
both are defined.

IX   __Energy__ $\int |f(t)|^2 dt = \int |\hat{f}(w)|^2 dw$ and more
generally

$$\int f(t)\overline{g(t)} \, dt = \int \hat{f}(w)\hat{g}(w) \, dw.$$

Moreover in the discrete case, the Fourier transform has
the best approximation property:

$$\int |f(t) - \Sigma_{|n|<N} \hat{f}(n)e^{2\pi i n t}|^2 dt$$

$$\leq \int |f(t) - \Sigma_{|n|<N} c_n e^{2\pi i n t}|^2 dt$$

with equality if and if $\hat{f}(n) = c_n$.

X   Special Cases

(a) $\hat{\delta}(w) = 1$ for all w.

(b) For an impulse response (or a signal, for that
     matter) h, its energy spectrum is $|\hat{h}|^2 = |H|^2$
     and the inverse Fourier transform, ρ, of the
     energy spectrum is the autocorrelation of h.
     From IV we have

$$\rho = h(t) * \overline{h}(-t)$$
$$= \int h(t-u)\, \overline{h}(-u)\, du$$
$$= \int h(t+v)\, \overline{h}(v)\, dv .$$

(c) A signal f is time-limited (TL) if f has finite
     energy and, for some $t_0 > 0$,

$$f(t) = 0 \quad \text{whenever} \quad |t| > t_0.$$

     (f is sometimes called a window signal.)

(d) A signal f is band-limited (BL) if its energy is
     finite and, for some $w_0 > 0$,

$$\hat{f}(w) = 0 \quad \text{whenever} \quad |w| > w_0.$$

     We have already seen the use of this concept in the
     example in section 3. An important consequence is
     that for any signal g,

$$\int g(t)f(t)e^{-2\pi iwt}\, dt = (f.g)\hat{}(w)$$
$$= (\hat{f} * \hat{g})\,(w).$$

     This is to be compared with V.

(e) An impulse train (with period 1) is

$$\mu(t) = \Sigma_{n=-\infty}^{\infty}\, \delta(t+n)$$

     and its transform is also an impulse train

$$\hat{\mu}(w) = 2\pi\, \Sigma_{n=-\infty}^{\infty}\, \delta(w-2\pi n).$$

(f) A normal signal has the shape

$$n(t) = e^{-\pi t^2}$$

     which is self-transform:

$$\hat{n}(w) = e^{-\pi w^2}.$$

5. APPLICATIONS

In this section we give simple applications of the previous ideas.

Circuit Analysis

An electric circuit with one input and one output can be regarded as a system; it is called <u>passive</u> if it is composed entirely of resistances, capacitances and inductances (i.e. no power amplifying devices like transistors). It can be shown that a circuit is passive if and only if, for any signal f,

$$\int (|f(t)|^2 - |(Tf)(t)|^2)dt \geqslant 0$$

which holds if and only if

$$1 - H(w)\overline{H(w)} \geqslant 0.$$

Furthermore it is lossless (i.e. a conservative system) if and only if

$$1 - H(w)\overline{H(w)} = 0$$

(see Wohlers, [9]).

Stability

A system is <u>externally stable</u> if a bounded signal yields a bounded response (sometimes called BIBO). A necessary and sufficient condition is that

$$\int |h(t)|dt < \infty$$

where h is the impulse response of the system.

Filters

It can be shown that if the transform $\hat{f}(w)$ of some signal f vanishes for $w < 0$ and if f vanishes in some interval (a,b) with $0 < a < b$ then the signal is zero. Consequently a filter, being frequency limited, does not

provide a causal system! Fortunately it can be approximated
by causal systems: these are "smoothed" versions of the original.   Indeed if H   is the system function of a (low-pass,
ideal) filter:

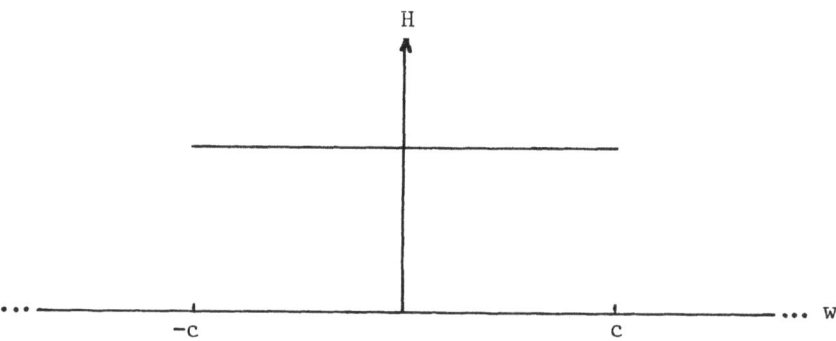

we've already seen that $h(t) = \sin(2\pi ct)/(\pi t)$. It remains
to take

$$h_a = t_a(hf_a) \qquad \text{for } a > 0$$

where f is the triangle signal

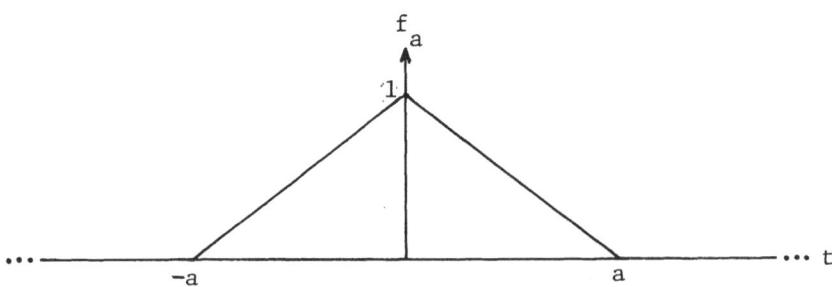

with transform

$$H_a(w) = e^{-2\pi i a w}.H(w) * (2\sin^2(aw/2)/(\pi a w^2))$$

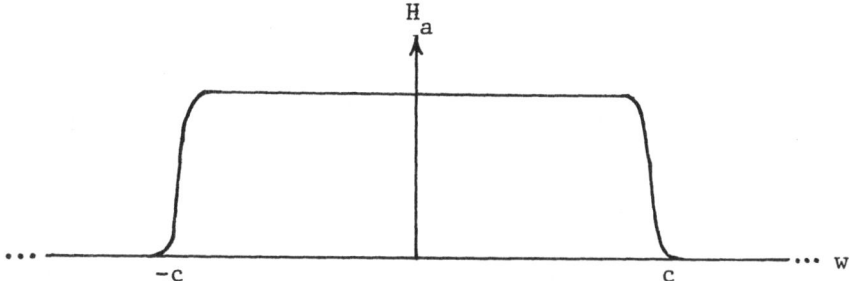

Now if we let $a \to \infty$, we find that $|H_a(w)| \to H(w)$ where H_a is causal.

Sampling

Often it is necessary to reconstruct an analogue signal by taking equally spaced samples from it. We can represent the sampled (discrete) signal as follows. Suppose the original (analogue) signal is $h(t)$

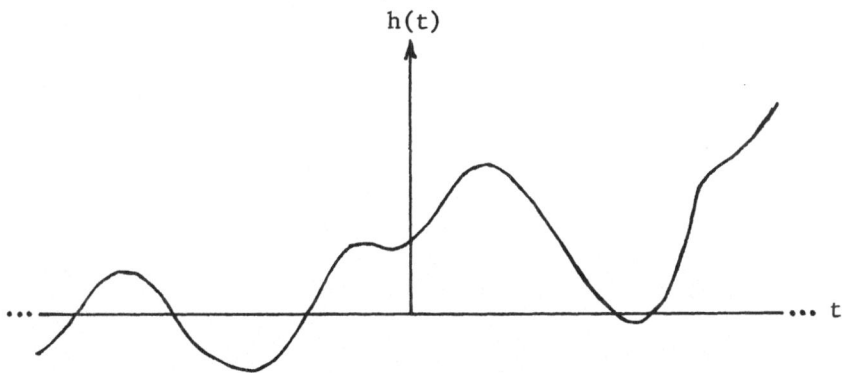

and that we write $\mu(t)$ for the impulse train whose period equals the distance apart of the samples

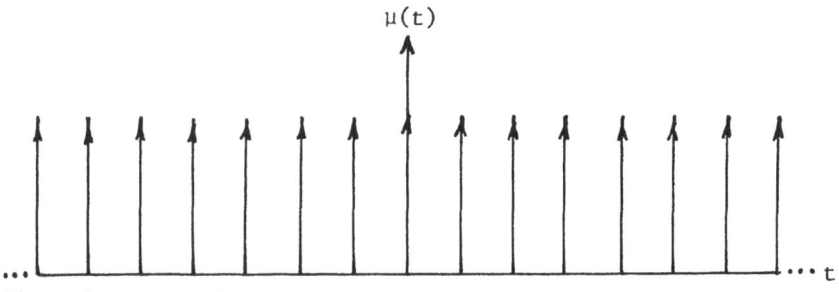

μ(t)

Then the sampled version is simply h(t)μ(t)

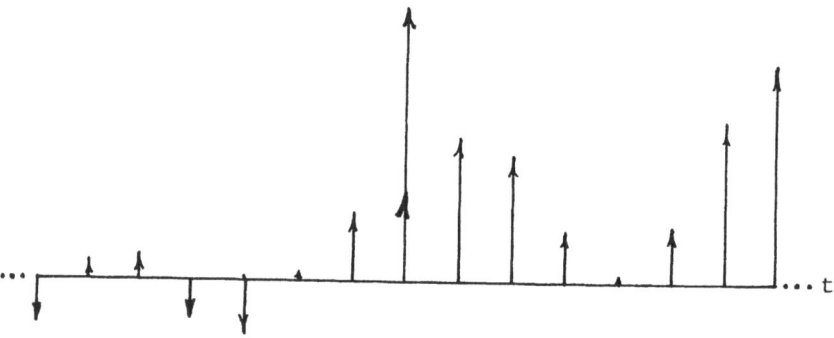

Now provided the original signal h is band limited we can
apply property X(d) to conclude that the Fourier transform
of h(t)μ(t) is (H*M)(w) where H and M are the Fourier
transforms of h and μ respectively

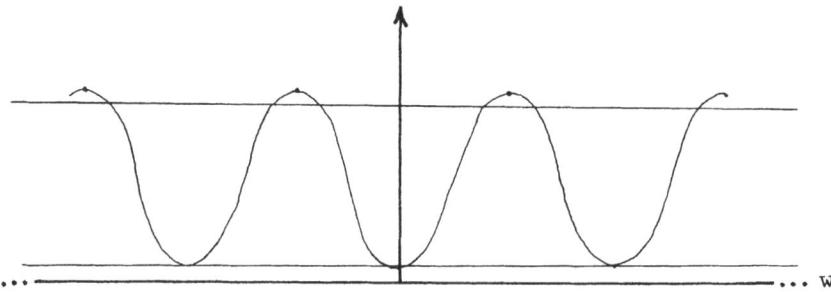

But this is the concatenation of H many times over (by
X(e)) so capturing one period of it we obtain H, and by
inverse transforming, finally obtain h.

There may be difficulties however. If the samples are distance L apart then the period of μ is L hence the period of M is L^{-1}. When h is BL with limit w_0 then we need

$$L < (2w_0)^{-1}.$$

Indeed if L is too large then we get distortion of $H*M$ due to "doubling up". This is called aliasing.

Computing

To multiply either polynomials or integers we use the same process, namely

$$(\Sigma_{j=0}^n \, a_j x^j)(\Sigma_{j=0}^n \, b_j x^j) = \Sigma_{j=0}^{2n} \, (\Sigma_{i=0}^n \, a_{j-i} b_i) x^j$$

which involves discrete convolution. This can be performed efficiently by transforming each multiplicand using the discrete Fourier transform (DFT), multiplying the transforms, then finding inverse transforms.

The DFT of an array $(a_0, a_1, \ldots, a_n)$ is given by the product with a matrix F whose entries are powers of a principal root of unity. Instead of $O(N^2)$ multiplications to evaluate the multiplication above, the fast Fourier transform (FFT) takes only $O(N \log N)$. This is known to be the fastest possible in certain cases, and explains the widespread use of the FFT in computing. For further details of the FFT algorithm, see [1] or [7].

A further use of the Fourier transform in computing is in testing a computer's random number generator for randomness by the spectral test.

REFERENCES

1. R. Bracewell, "The Fourier Transform," McGraw Hill, London (1965).

2. H. Dym and H.P. McKean, "Fourier Series and Integrals," Academic Press, New York (1972).

3. R.E. Edwards, "Fourier Series: A modern Introduction," Volume 1, Second Edition, Springer-Verlag, Berlin (1979).

4. W. Feller, "An Introduction to Probability Theory and
    Its Applications," Volume 1 (Third Edition), 1968 and
    Volume 2 (Second Edition), 1971, John Wiley and Sons,
    New York.

5. T. Kailath, "Linear Systems," Prentice-Hall, Englewood
    Cliffs (1980).

6. D.E. Knuth, "The Art of Computer Programming: Volume 1,
    Fundamental Algorithms," Second Edition, Addison-Wesley,
    Reading (1968).

7. A. Papoulis, "The Fourier Integral and Its Applica-
    tions," McGraw-Hill, London (1962).

8. A. Papoulis, "Signal Analysis," McGraw-Hill, London
    (1977).

9. M.R. Wohlers, "Lumped and Distributed Passive Networks.
    A Generalized and Advanced Viewpoint," Academic Press,
    New York (1969).

UNCERTAINTY PRINCIPLES AND SAMPLING THEOREMS

John F. Price

School of Mathematics
University of New South Wales
Kensington, N.S.W., Australia

In this lecture I will be dealing with two separate topics both of which are important for a wide range of applications of Fourier analysis. The first is that of uncertainty. Roughly speaking, it says that the more we concentrate a signal in time, the greater is its bandwidth, and vice versa. (The first explicit statement of this reciprocity is in the mathematics of the quantum mechanical uncertainty principle of Heisenberg, and so the name "uncertainty" is usually applied to all results of this nature.)

The second topic is the sampling theorem; it provides a method for reconstructing a bandlimited signal from its values (or samples) at certain evenly spaced points in time. This section can be used as a lead-in to Henry Landau´s presentation "An overview of time and frequency limiting" given later.

1. RECIPROCITY IN FOURIER ANALYSIS

There are many senses in which there is a reciprocity or duality between a signal or function and its Fourier transform. One of the most useful of these is based on the fact that if F is the Fourier transform of f, then (up to a constant) $\omega F(\omega)$ is the Fourier transform of its derivative f´. This, in effect, means there is a duality between differentiability properties of f and algebraic properties of F.

25

Another fundamental duality is between the duration of a signal and its bandwidth. The general idea is that the more we concentrate a signal in time, the more spread out is its Fourier transform, and vice versa. This ubiquitous principle has been given precision in a number of ways. The oldest of these is the well-known result that we cannot have a nonzero signal which is both time-limited and band-limited. The general theme of the first part of the lecture is an examination of how close we can come to constructing such a signal. Putting this in another way, given a family of signals of some specified duration or spread, what is the minimum bandwidth which can be realized by such signals?

There are numerous answers to this, but before looking at some of them, we introduce a small amount of notation.

The Lebesgue spaces L^p, where $1 \leqslant p < \infty$, are defined as the space of (measurable) functions f for which $\|f\|_p < \infty$ where

$$\|f\|_p = \begin{cases} (\int |f(t)|^p dt)^{1/p} & \text{if } 1 \leqslant p < \infty \\ \text{ess sup}\{|f(t)| : t \in R\} & \text{if } p = \infty. \end{cases}$$

(Unless stated otherwise, all integrals are over the real line $\mathbb{R}$. The idea of the "essential supremum" is that it restricts membership of L^∞ to functions bounded everywhere except possibly on sets of measure zero.)

Given $f \in L^1$, its Fourier transform F or $\hat{f}$ is defined by

$$F(\omega) = \hat{f}(\omega) = \int f(t) e^{-2\pi i \omega t} dt \qquad (\omega \in R).$$

Evidently $\|F\|_\infty \leqslant \|f\|_1$.

It may be shown that $\|f\|_2 = \|F\|_2$ for functions $f \in L^1 \cap L^2$ and since these functions are dense in L^2 it is natural to extend the definition of Fourier transform to all of L^2. When this is done we obtain Plancherel's formula:

$$\int f(t)\overline{g(t)} = \int F(\omega)\overline{G(\omega)} d\omega,$$

valid for all $f,g \in L^2$. (The space L^2 is actually a Hilbert space with an inner product defined by

$$(f,g) = \int f(t)\overline{g(t)}dt.)$$

The $\underline{energy}$ of f is defined as

$$E_f = \int |f(t)|^2 dt = \int |F(\omega)|^2 d\omega.$$

2. REVIEW OF UNCERTAINTY PRINCIPLES

$\underline{2.1}$ $\underline{Heisenberg}$ $\underline{uncertainty}$ $\underline{principle}$. The mathematical core of Heisenberg's uncertainty principle in quantum mechanics established in 1927 is the inequality

$$\int t^2 |f(t)|^2 dt \cdot \int \omega^2 |F(\omega)|^2 d\omega \geqslant (16\pi^2)^{-1} E_f^2. \qquad (2.1)$$

Clearly $\int t^2 |f(t)|^2 dt / E_f = (\|tf\|_2 / \|f\|_2)^2$ is a measure of the spread of the signal f, and similarly for F. Hence (2.1) indicates that the more f is concentrated (as measured in this way), the greater the spread of F, and vice versa.

The first use of (2.1) outside quantum mechanics was made by Stewart [24] in 1931. In 1946 Gabor [8] used the inequality to give a precise meaning to the piece of folk- lore that the information conveyed by a frequency band in a given time interval has an upper bound. (A fascinating dis- cussion of this theme is contained in Slepian [23].)

As will be shown below, equality occurs in (2.1) if and only if $f(t) = \text{const.}\exp(-kt^2)$ for some $k > 0$. Hence (2.1) is sharpest when the signal is approximately of this form, that is, approximately a constant multiple of a normal signal. (A signal f is said to be $\underline{normal}$ or $\underline{Gaussian}$ if it is of the form

$$f(t) = (\sqrt{2}\pi\sigma)^{-1} \exp(-t^2/2\sigma^2) \quad \text{for some } \sigma > 0.)$$

On the other hand, when signals are considered which are further and further away from normal, the inequality becomes less and less useful. For example, if $f(t) \sim |t|^\alpha$ as $|t| \to \infty$ where $-3/2 \leqslant \alpha < -1/2$, then $E_f < \infty$ but $\|tf\|_2 = \infty$, or if f is a square wave, $\|\omega F\|_2 / E_f = \infty$. Thus (2.1) gives no usable estimates in these cases.

<u>Theorem.</u>   For all signals f with finite energy,

$$\int t^2 |f(t)|^2 dt \cdot \int \omega^2 |F(\omega)|^2 d\omega \ge (16\pi^2)^{-1} E_f^2$$

with equality if and only if $f(t) = const.exp(-kt^2)$ for
some $k > 0$.

<u>Proof.</u>   Let E be the set of complex-valued continuously
differentiable functions f on $\mathbb{R}$ for which $t|f(t) \to 0$
as $t \to \infty$. Assume for the moment that $f \varepsilon E$. The familiar
formula

$$(f')\hat{}(\omega) = 2\pi i\omega \, F(\omega)$$

is obtained using integration by parts. Also it is clear
that

$$2|\overline{f}f'| \ge \overline{f}f' + f\overline{f'} = |f^2|'.$$

Combining these with Plancherel's formula and Schwartz's inequality gives

$$16\pi^2 \int |tf(t)|^2 dt \int |\omega F(\omega)|^2 d\omega = 4\int |tf(t)|^2 dt \int |(f')\hat{}(\omega)|^2 d\omega$$
$$= 4\int |tf(t)|^2 dt \int |f'(t)|^2 dt$$
$$\ge 4(\int |t\overline{f}(t)f'(t)|dt)^2$$
$$\ge (\int td|f(t)|^2)^2$$
$$= \int |f(t)|^2 \, dt \, ,$$

as asserted.

Now suppose that f has finite energy and that both
$\int |tf(t)|^2 dt$ and $\int |\omega F(\omega)|^2 d\omega$ are finite. (If they are not,
then there is nothing to prove.) Standard techniques show
how to choose a sequence (f_n) in E (we may even suppose
that all the f_n's have bounded support) so that

$$\lim_{n \to \infty} \int |f(t) - f_n(t)|^2 dt \, ,$$

$$\lim_{n \to \infty} \int t^2 |f(t) - f_n(t)|^2 dt = 0$$

and $$\lim_{n \to \infty} \int \omega^2 |\hat{F}(\omega) - \hat{F}_n(\omega)|^2 d\omega = 0.$$

Putting $f = f_n$ in the inequality already established and letting $n \to \infty$ gives the required inequality.

Now suppose that we have equality with $f \neq 0$. In the first place this means that $2|ff'| = ff' + ff'$. This can only happen if $re(f)$ and $im(f)$ are multiples of each other. Hence suppose that $f = c\phi$ where $c \in \mathbb{C}$ and ϕ is a real-valued function. Equality in the use of the Schwartz inequality requires $f' = const.tf(t)$ which means

$$\phi'(t) = kt\phi(t)$$

for some $k \in \mathbb{R}$. The only solutions of this differential equation which have finite energy are multiples of $\exp(kt^2/2)$ where $k < 0$. Hence these are the only functions for which we have equality in the Heisenberg uncertainty principle inequality. (Strictly speaking we have only shown that these are the only functions in E for which we have equality. One has to work a little harder to obtain this aspect of the result when all functions with finite energy are allowed.)

2.2 Hardy's Theorem.

Suppose we have a signal satisfying

$$|f(t)| \leqslant const. \exp(-\alpha t^2) \qquad (t \in R)$$

$$|F(\omega)| \leqslant const.\exp(-\beta \omega^2) \qquad (\omega \in R)$$

In 1933 Hardy [9] showed that $f = 0$, or f is a constant multiple of $\exp(-\alpha t^2)$, or there are infinitely many such functions f according as $\alpha\beta > \pi^2$, $\alpha\beta = \pi^2$, or $\alpha\beta < \pi^2$. Hence it is impossible to construct a nonzero signal f which, along with its Fourier transform is "rapidly decreasing" at infinity. Extensions of this result are contained in [5].

2.3 Entropy Considerations.

Following Shannon [22], the entropy of a signal f is defined as

$$H(f) = -\int |f(t)| \log|f(t)| dt ,$$

and similarly for its Fourier transform. In 1957 Hirschman [10] proved that

$$H(f) + H(F) \geqslant 1 - \log 2 \qquad (2.2)$$

for signals f with energy $E_f = 1$ (The above statement of the inequality makes use of a later result due to Beckner [1].)

Since entropy corresponds to information (and also, as it turns out, to concentration), (2.2) says that there is an upper limit to the total amount of information contained in a signal and its spectrum. A consequence of (2.2) given in [10] is a generalisation of (2.1), namely that

$$\left\| \, |t|^\alpha f \right\|_2 \cdot \left\| \, |\omega|^\alpha F \right\|_2 \geqslant H_\alpha E_f^2 \qquad (2.3)$$

for $\alpha > 0$, where $H_\alpha = 2\alpha e(8/e)^\alpha (\Gamma(1/2\alpha)/2\alpha)^{2\alpha}$.

2.4 The "Bell" Theorems.   Commencing in 1961, Landau, Pollak and Slepian of Bell Laboratories proved a series of results which took a direct approach to the problem of deciding how close a signal and its spectrum can be to being supported on finite intervals. Given T, Ω, define

$$\alpha^2 = \int_{|t| \leqslant T} |f(t)|^2 dt/E_f, \qquad \beta^2 = \int_{|\omega| \leqslant \Omega} |F(\omega)|^2 d\omega/E_f.$$

Clearly $0 \leqslant \alpha \leqslant 1$ and the closer α is to 1, the nearer f is to being concentrated on the interval $[-T,T]$. Similarly for F, β and Ω.

It is impossible for both $\alpha = 1$ and $\beta = 1$ for nonzero signals. To see this, suppose that $\alpha = 1$. Then $\hat{f}$ extends to an entire function in the complex plane. Hence if also $\beta = 1$, we must have $f = 0$. (This is simply the classical result mentioned in the introduction that nonzero functions cannot be both time-limited and band-limited.)

The above-mentioned authors determine a function $\gamma: \mathbb{R}^+ \to [0,1]$ with the property that the only possible pairs (α,β) lie in the subset of the square $[0,1] \times [0,1]$ described by

$$\cos^{-1}\alpha + \cos^{-1}\beta \geqslant \cos^{-1}(\gamma(T\Omega)^{1/2}),$$

except that the points $(0,1)$ and $(1,0)$ are excluded. In particular, if $\alpha = 1$, then $\beta \leqslant \gamma(T\Omega)^{1/2}$.

The function γ can be defined as follows. If nonzero f satisfies $\alpha = 1$, then

$$\int_{|\omega| \leqslant \Omega} |F(\Omega)|^2 d\omega < \int |f(t)|^2 dt$$

(otherwise f would also have $\beta = 1$). Now quantify this inequality by defining $\gamma(T\Omega)$ as the smallest number for which

$$\int_{|\omega|\leqslant\Omega}|F(\omega)|^2 d\omega)^{1/2} \leqslant \gamma(T\Omega)(\int |f(t)|^2 dt)$$

for nonzero f with $\alpha = 1$.

Further details of these results are given in the accompanying article by Henry Landau.

2.5 **Further Results.**   Examples of other uncertainty results are contained in Fuchs [7], Pearl [16], Benedetto [2] and Cowling and Price [4,5]. The inequalities in the latter two papers involve the norms $\||t|^\theta f\|_p$ and $\||\omega|^\phi F\|_q$ as measures of the "spread" of f and F respectively. Because the results are valid for a wide range of $\theta, \phi > 0$ and $p,q \in [1,\infty]$, the restriction which applies to (2.1) of needing f and F to decrease fairly rapidly at infinity is removed. Roughly speaking, any f with finite energy can be meaningfully accommodated within at least one inequality.

3. LOCAL UNCERTAINTY PRINCIPLES

If we concentrate more and more the duration of a signal while keeping its energy fixed, the results of the preceding section show that its bandwidth must increase. However, with the exception of 2.4, none of the inequalities rule out the possibility of this taking place by the spectrum forming two or more peaks far out from zero and remaining very small elsewhere. Local uncertainty principles show that if a signal is concentrated in time, then not only is its bandwidth large, but also its spectrum cannot be localized, that is, it cannot "peak" too much around any point. (Information of this type is also contained in the results of 2.4, but is somewhat obscured by the complexity of the function γ.)

A number of inequalities describing local uncertainty principles are given in Faris [6]. A typical result is:

3.1 **Theorem**   For all numbers a,b with $b > 0$ and signals f with finite energy,

$$\int_{a-b}^{a+b} |F(\omega)|^2 \, d\omega \leqslant 4\pi b \, |tf|_2 E_f^{1/2}.$$

Discussion.   One method of interpreting this inequality is
to first suppose that a,b are fixed. Then, as we consider
signals f which are increasingly concentrated about the
origin, that is signals for which $\|tf\|_2$ becomes smaller
and smaller, we see that the spectral energy in the band
[a-b, a+b] also decreases.

The proof given in Faris involves the notion of an
approximate identity. However, Henry Landau has pointed out
a simpler argument which gives the general form of the inequality (but with twice the constant). Suppose f is band-
limited and that its Fourier transform F has a continuous
first derivative. Since

$$2|\overline{F}F'| \geqslant \overline{F}F' + \overline{F'}F = |F^2|' \, ,$$

Schwartz's inequality shows that

$$|F^2|(\omega) = \int_{-\infty}^{\omega} |F^2|'(y)dy$$
$$\leqslant 2\int |\overline{F}F'|dy \leqslant 2\|F\|_2\|F'\|_2$$

for each $\omega \varepsilon \mathbb{R}$. By Parseval's formula, as well as integra-
tion by parts in the second case, $\|F\|_2 = E_f^{1/2}$ and
$\|F'\|_2 = 2\pi\|tf\|_2$ so that

$$\int_{a-b}^{a+b} |F(\omega)|^2 d\omega \leqslant \int_{a-b}^{a+b} 4\pi \, E_f^{1/2} \, |tf|_2 dy$$

$$= 8\pi b \, E_f^{1/2}|tf|_2.$$

A limit argument similar to that in the proof of Theorem 2.1
completes the proof (with a constant $8\pi b$ instead of $4\pi b$).

The inequality in Theorem 3.1 suffers from the difficulty mentioned in 2.1 and 2.5, namely that f needs to
decrease fairly rapidly to zero as $|t| \to \infty$. Families of ine-
qualities are given in Price [18,26] which allow a range of
powers of t in the expression on the right-side. Below are
two examples of these results. (Γ is the usual gamma function defined by

$$\Gamma(u) = \int_0^{\infty} e^{-t}t^{u-1}dt$$

for u > 0. Recall that $\Gamma(1/2)^2 = \pi$ so that, apart from the strict inequality, Theorem 3.3 reduces to Theorem 3.1 when $\alpha = 1$.)

3.2 Theorem [18]. Suppose $0 \leqslant \theta < 1/2$. For all numbers a,b with $b > 0$ and signals f,

$$(\int_{a-b}^{a+b} |F(\omega)|^2 d\omega)^{1/2} \leqslant (1+(\frac{2}{1-2\theta})^{1/2})b^\theta \||t|^\theta f\|_2.$$

For θ outside this range no such inequality is possible.

3.3 Theorem [26]. Suppose $\alpha > 1/2$ and a,b are numbers with $b > 0$. For all signals f

$$\int_{a-b}^{a+b} |F(\omega)|^2 d\omega < Kb\|f\|_2^{2-1/\alpha} \||t|^\alpha f\|_2^{1/\alpha}$$

where

$$K = \frac{2}{\alpha} \Gamma(\frac{1}{2\alpha})\Gamma(1-\frac{1}{2\alpha})(2\alpha-1)^{1/2\alpha}(1-\frac{1}{2\alpha})^{-1}$$

Moreover, Kb is the smallest possible constant.

The proofs of these results are given in [18] and [26]. In the former paper the inequalities are used to estimate quantum mechanical Hamiltonians. In [20], a family of local uncertainty inequalities is obtained for periodic signals.

4. APPLICATIONS

4.1 Quantum Mechanics.   In quantum mechanics it is usual to normalise f so that $E_f = 1$; as such it represents the state of a one-dimensional system. Proceeding with this interpretation, $\int t^2 |f(t)|^2 dt$ and $\int \omega^2 |F(\omega)|^2 d\omega$ represent the variances of the position and momentum observables respectively (assuming they both have mean zero). In this way, inequality (2.1) becomes the mathematical formulation of the quantum mechanical uncertainty principle first described by Heisenberg in 1927. It simply says that the smaller the variance of position (that is, the less the "uncertainty"

regarding the position of, say, a particle), the greater the
variance of momentum (that is, the less we can be certain of
the momentum of the particle.)

Applications of local uncertainty principles to quantum
mechanics are given in [6] and [18].

4.2 Spectrum Concentration.   The inequalities of 3 can be
used to obtain estimates for the percentage of the energy of
a specified signal to be in a given band. The problems can
be formulated as follows: given $0 < c < 1$, choose $b > 0$
so that

$$\text{(A)} \quad \int_{-b}^{b} |F(\omega)|^2 d\omega > c\, E_f \,, \quad \text{or}$$

$$\text{(B)} \quad \int_{-b}^{b} |F(\omega)|^2 d\omega \leqslant c\, E_f.$$

Estimate (A).   The inequality (A) is valid for any b
satisfying

$$4\pi^2 (1-c) b^2 E_f \geqslant \int |f'(t)|^2 dt. \qquad (4.1)$$

Proof.   Suppose b satisfies (4.1). Then

$$\int_{|\omega| \geqslant b} |F(\omega)|^2 d\omega \leqslant b^{-2} \int \omega^2 |F(\omega)|^2 d\omega$$

$$= b^{-2} (2\pi)^{-2} \int |(f')\hat{\,}|^2 d\omega$$

$$= (2\pi b)^{-2} \int |f'(t)|^2 dt$$

$$\leqslant (1-c) E_f \,,$$

whence $\int_{|\omega| \leqslant b} |F(\omega)|^2 d\omega \geqslant c\, E_f$, as required.

Estimate (B).   The inequality (B) is valid for any b
satisfying

$$(1+(2/(1-2\theta)))^{1/2})b^{\theta} \; ||t|^{\theta}f|_2 \; < \; (c \; E_f)^{1/2},$$

where $0 < \theta < 1/2$, and for any b satisfying

$$b \; < \; K^{-1}E_f^{1/2\alpha}||t|^{\theta}f|_2^{-1/\alpha}$$

where $\theta > 1/2$ and K is as defined in 3.3 above.

Proof.   An immediate consequence of 3.2 and 3.3.

5. THE SHANNON SAMPLING THEOREM

The classical sampling theorem provides a method for reconstructing a signal from its values at certain evenly spaced points in time, the spacing being given by the so-called Nyquist rate. An immediate consequence is that the more restricted the bandwidth, the slower (or more expanded) is the Nyquist rate, and vice versa.

Key names associated with the development of the sampling theorem are Cauchy [3], Whittaker [25], Nyquist [15], Kotel'nikov [13] and Shannon [21]. It is now usually referred to as the Shannon sampling theorem since it was he who brought out its usefulness in information theory. A brief history of the result is given in Jerri [11] as a prelude to a comprehensive discussion of the result and its applications. (This review article concludes with nearly 250 references.)

The result states that given $\Omega > 0$,

$$f(t) \; = \; \Sigma_{n \in \mathbb{Z}} f(n/2\Omega) \frac{\sin 2\pi\Omega(t-n/2\Omega)}{2\pi\Omega(t-n/2\Omega)} \tag{5.1}$$

for all continuous functions f with finite energy such that $F(\omega) = 0$ for $|\omega| > \Omega$. In other words, such a function f can be reconstructed by knowing its values at the points $\{n/2\Omega : n \in \mathbb{Z}\}$. We will state and prove a generalisation of this result due independently to Petersen and Middleton [17] and to Kluvánek [12].

5.1 Definition.   A (measurable) subset E of $\mathbb{R}$ is said to be an Ω - transversal set if

$$\Sigma_{n\varepsilon Z} \ \chi_E(\omega+2\Omega n) = 1$$

for all $\omega \ \varepsilon \ \mathbb{R}$ where χ_E is the characteristic function of E.

It is easy to see that E being an Ω-transversal set means simply that for each $\omega \ \varepsilon \ \mathbb{R}$ the set of points $\{\ldots,\omega-2\Omega,\omega,\omega+2\Omega,\ldots\}$ intersects E at precisely one point. The canonical example of these sets is $(-\Omega,\Omega]$. Since

$$\int \chi_E(\omega)d\omega = \int_{-\Omega}^{\Omega} \Sigma_{n\varepsilon Z} \ \chi_E \ (\omega+2\Omega n)d\omega,$$

it follows that the measure of every Ω-transversal set is 2Ω.

Given $E \subseteq \mathbb{R}$ with finite measure, a key function in the statement of the theorem is $\psi_E = \mathbb{F}^{-1}\chi_E/2\Omega$. In other words,

$$\psi_E(t) = (2\Omega)^{-1} \int_E e^{2\pi i\omega t} d\omega. \qquad (5.2)$$

Notice that when $E = (-\Omega,\Omega]$,

$$\psi_E(t) = (2\pi\Omega t)^{-1}\sin 2\pi\Omega t \qquad (5.3)$$

which, along with its translates by $n/2\Omega$, make up the functions on the right side of (5.1).

Let L_E^2 denote the space of function in L^2 whose Fourier transforms have support in E. (The support of a function is the closure of the set of points on which it is nonzero.)

In view of (5.3), evidently the following result is a generalisation of the classical sampling theorem described in (5.1).

5.2 Theorem.   Let E be an Ω-transversal set.   Then

$$f(t) = \Sigma_{n\varepsilon Z}f(n/2\Omega)\psi_E(t-n/2\Omega) \qquad (t \ \varepsilon \ \mathbb{R}) \qquad (5.4)$$

for all continuous $f \ \varepsilon \ L_E^2$ where the series converges in the L^2-sense and uniformly.

Proof.   For each $n \ \varepsilon \ \mathbb{Z}$, define $\phi_n(t) = (2\Omega)^{1/2}\psi_E(t-n/2\Omega)$

and assume for the moment the following two facts:

$$(\phi_n)_{n \epsilon Z} \text{ is a complete orthonormal family in } L_E^2, \quad (5.5)$$

$$(f,\phi_n) = (2\Omega)^{-1/2} f(n/2\Omega) \quad \text{for continuous } f \epsilon L_E^2. \quad (5.6)$$

Since L_E^2 is a Hilbert space, (5.3) implies that for each $f \epsilon L_E^2$ there exist complex number $(c_n)_{n \epsilon Z}$ such that

$$f = \Sigma_{n \epsilon Z} c_n \phi_n \qquad \text{in } L_E^2, \qquad (5.7)$$

where $c_n = (f,\phi_n)$. If f is also continuous,

$$c_n = (2\Omega)^{-1/2} f(m/2\Omega) \quad \text{by (5.6)}.$$

Hence (5.4) is valid in the L^2-sense. Now for $g \epsilon L_E^2$,

$$\|g\|_\infty \leqslant \|\hat{g}\|_1 \leqslant (2\Omega)^{1/2} \|\hat{g}\|_2 = (2\Omega)^{1/2} \|g\|_2,$$

where the first inequality is stated in the introduction and the second follows by the Schwartz inequality. This means that for continuous f in L_E^2, L^2-convergence implies uniform convergence and so (5.4) is attained.

It remains to prove (5.5) and (5.6). From the definition (5.2) of ψ_E, it follows that

$$\hat{\phi}_n(\omega) = (2\Omega)^{-1/2} \exp(-\pi i n\omega\Omega) \chi_E(\omega).$$

Hence by Plancherel's identity, whenever $f \epsilon L_E^2$,

$$(f,\phi_n) = (2\Omega)^{-1/2} \int_E F(\omega) \exp(\pi i n\omega/\Omega) d\omega \qquad (5.8)$$

so that if f is also continuous we have (5.6) as required.

The proof of orthonormality of the family (ϕ_n) also begins with an application of Plancherel's identity:

$$(\phi_m,\phi_n) = (2\Omega)^{-1} \int \chi_E(\omega) \exp(\pi i (n-m)\omega/\Omega) d\omega$$

$$= (2\Omega)^{-1} \Sigma_{k\epsilon \mathbf{Z}} \int_{-\Omega}^{\Omega} \chi_E(\omega+2k\Omega) \exp(\pi i(n-m)(\omega+2k\Omega)/\Omega) d\omega$$

$$= (2\Omega)^{-1} \int_{-\Omega}^{\Omega} \Sigma_{k\epsilon \mathbf{Z}} \chi_E(\omega+2k\Omega) \exp(\pi i(n-m)(\omega+2k\Omega)/\Omega) d\omega$$

$$= (2\Omega)^{-1} \int_{-\Omega}^{\Omega} \exp(\pi i(n-m)\omega/\Omega) d\omega = \delta_{mn},$$

where the penultimate equality is a consequence of the Ω-transversality of E.

Finally we prove that the family (ϕ_n) is complete in L_E^2. Given $f \epsilon L_E^2$ with $(f,\phi_n) = 0$ for all n, with little effort we observe from (5.8) that

$$\int_{-\Omega}^{\Omega} \Sigma_{k\epsilon \mathbf{Z}} F(\omega+2k\Omega) \exp(\pi i n\omega/\Omega) d\omega = 0 \quad \text{for all } n.$$

(Make use of the Ω-transversality of E as in the previous paragraph.) This means that all the Fourier coefficients of $\Sigma_{k\epsilon \mathbf{Z}} f(\omega+2k\Omega)$, considered as a function on $[-\Omega,\Omega]$, are zero. As such this function is square integrable and so equals zero almost everywhere. Since supp $F \subseteq E$ and E is Ω-transversal, $F = 0$ in $L^2(E)$, whence $f = 0$ in L_E^2, as required.

5.3 Higher Dimensions. There are straightforward generalisations of both (5.1) and (5.4) to higher dimensions. However, in this case it is a considerably more delicate and interesting pursuit to obtain the most economical spacing (that is, the most open spacing) of the sample points. This problem is examined in [17] and [19]. In the latter paper it is also shown that, roughly speaking, (5.4) with E an Ω-transversal set is the only way of reconstructing continuous functions $f \epsilon L_E^2$ from their values on $\{n/2\Omega : n \epsilon \mathbf{Z}\}$.

A consequence of standard L^2-theory applied to (5.5) and (5.6) is:

5.4 Corollary. Whenever (continuous) f belongs to L_E^2, where E is an Ω-transversal set,

$$\|f\|_2 = ((2\Omega)^{-1} \Sigma_{n\epsilon \mathbf{Z}} |f(n/2\Omega)|^2)^{1/2}.$$

5.5 Applications.   Apart from information theory, the area
in which Shannon first introduced the result, the sampling
theorem has found application in such areas as optics, crystallography, meteorology and boundary value problems.
Detailed references are given in [11]. Here we give only one
application, namely the way that it can be used to give precision to the price of folk-lore that there are approximately
$4\Omega T$ linearly independent signals f satisfying supp f $\subseteq$
$[-T,T]$ and   supp F $\subseteq [-\Omega,\Omega]$. Of course, as it stands it is
false since f = 0 is the only such signal. Nevertheless it
is approximately true, as we shall see, in a quite specific
sense. A deeper analysis of this result is given in [23].

Whenever a $\varepsilon \mathbb{R}$, define $f_a(t) = f(t+a)$. For simpli-
city assume that $2\Omega T$ is integral.

5.6 Corollary.   Given $\Omega T > 0$, let E be an Ω-
transversal set. For each continuous f in L_E^2 there
exists a in $[0,1/2\Omega]$ so that

$$\|f_a - \Sigma_{|n| \leqslant 2\Omega T} f_a(n/2\Omega)\psi_E(t-n/2\Omega)\| \leqslant (\int_{|t|>T}|f(t)|^2 dt)^{1/2}.$$

Discussion.   Let $V_{\Omega T}$ be the $(4\Omega T+1)$-dimensional subspace of L_E^2 generated by the functions $\psi_E(t-n/2\Omega)$ where
$|n| \leqslant 2\Omega T$. The interpretation of the preceding result is
that given continuous f in L_E^2, there exists a "small"
time lag a with $0 \leqslant a \leqslant 1/2\Omega$ such that f_a "almost"
belongs to $V_{\Omega T}$. In terms of energy, the amount by which it
misses out is bounded by

$$\int_{|t|>T}|f(t)|^2 dt.$$

Hence as f becomes increasingly concentrated in $[-T,T]$,
this discrepancy becomes smaller.

Proof (of 5.6). With T,Ω,E and f as in the statement of
the corollary

$$\int_0^{1/2\Omega} \|f_a - \Sigma_{|n|<2\Omega T} f_a(n/2\Omega)psy_E(t-n/2\Omega)\|_2^2 \, da$$

$$= \int_0^{1/2\Omega}(2\Omega)^{-1} \Sigma_{|n|>2\Omega T}|f_a(n/2\Omega)|^2 \, da$$

$$\leqslant (2\Omega)^{-1} \int_{|t|>T} |f(t)|^2 \, dt,$$

where the first equality follows from 5.4. The observation
that if

$$\int_0^{1/2\Omega} |\phi(a)| \, da \leqslant k, \quad \text{then} \quad |\phi(a)| \leqslant 2\Omega k$$

for some a in $[0,1/2\Omega]$ completes the proof.

6.   APPROXIMATIONS VIA THE SAMPLING REPRESENTATION

In this section we give a very brief survey of some of
the types of errors which may occur when the sampling theorem
is applied in practical situation. For further details the
reader is invited to consult [11] and the numerous references
given there.

6.1 Truncation error.   In practice a given signal f can
only be sampled at a finite number of points – the resulting
error is called the truncation error. A simple estimate of
this error is given by the following result.

Theorem.   Let $E \subseteq \mathbb{R}$ be an Ω-transversal set. Given (continuous) f in L_E^2 and n in $\{1,2,\ldots\}$,

$$|f(t) - \Sigma_{|n| \leqslant N} f(n/2\Omega) \psi_E(t-n/2\Omega)| \qquad (6.1)$$

$$\leqslant \|f\|_2 (2\Omega \Sigma_{|n|>N} |\psi_E(t+n/2\Omega)|^2)^{1/2}$$

for all t in $\mathbb{R}$.

Proof.   Given t in $\mathbb{R}$ and N in $\{1,2,\ldots\}$, define
$$\Psi_{N,t}(\omega) = [\exp(2\pi i t\omega) - \Sigma_{|n| \leqslant N} \exp(\pi i n\omega/\Omega)\psi_E(t-n/2\Omega)]\chi_E(\omega).$$
Since $F^{-1}(\exp(2\pi i a\omega)F(\omega))(0) = f(a)$, it follows that for
continuous f in L_E^2,

$$f(t) - \Sigma_{|n| < N} f(n/2\Omega) \psi_E(t-n/2\Omega) = F^{-1}(\Phi_{N,t}F)(0).$$

Consequently the left side of (6.1) satisfies

$$|\ldots| \leqslant \|F^{-1}(\Phi_{N,t}F)\|_\infty \leqslant \|\Phi_{N,t}\|_2 \|F\|_2.$$

It now remains to find the appropriate bound for $\|\Phi_{N,t}\|_2$. From (5.5) in the preceding section we know that the family $(\phi_n)_{n \in Z}$ is complete and orthonormal in $L^2(E)$. Also

$$(\exp(2\pi i t\omega), \hat{\phi}_n) = \phi_{-n}(t) = (2\Omega)^{1/2} \psi_E(t+n/2\Omega).$$

Hence the sum in the definition of $\Phi_{N,t}$ is the beginning of the expansion of the first term of $\hat{\phi}_{N,t}$ with respect to the orthonormal basis $(\hat{\phi}_n)$. Consequently

$$\|\Phi_{N,t}\|_2 = (\Sigma_{|n|>N} |(\exp(2\pi i t\omega), \hat{\phi}_n)|^2)^{1/2}$$

$$= (2\Omega \; \Sigma_{|n|>N} \; |\psi_E(t+n/2\Omega)|^2)1/2,$$

as required.

6.2 <u>Aliasing</u> <u>Error</u>.   It is frequently the case that the signal f to be considered is not bandlimited or, if it is bandlimited, its bandwidth is inconveniently large. If we proceed as if the signal f is bandlimited or, more generally, that it belongs to L_E^2 where E is an Ω-transversal set, the resulting discrepancy is referred to as the <u>aliasing</u> <u>error</u>.   (It is such a truncation or restriction of the spectrum which makes wagon wheels often appear to go backwards in Western films. Here the signals are the positions of the spokes and the sequence of still photographs which make up the film have the effect of restricting their Fourier transforms.)

The basic inequality for aliasing errors arises in the following manner. Given (measurable) $E \subseteq \mathbb{R}$, define f_E so that its Fourier transform is $\chi_E F$. Then $f_E \in L_E^2$ and if also $F \in L^1$,

$$|f(t) - f_E(t)| \leqslant \int_{\omega \in E} |F(\omega)| d\omega \qquad \text{for } t \in \mathbb{R}.$$

This is established by the string of inequalities:

$$|f(t) - f_E(t)| \leqslant \|f-f_E\|_\infty \leqslant \|f-f_E\|_1 .$$

<u>6.3 Jitter Error.</u>   Suppose that the sampling formula
implies that a signal f can be recovered by sampling at the
points $\{t_n : n \in \mathbb{Z}\}$. If instead, for whatever reason, the
signal is sampled at the points $\{t_n + \varepsilon_n : n \in \mathbb{Z}\}$, where the
ε_n satisfy some deterministic or statistical pattern, the
resulting error is called the <u>jitter</u> <u>error</u>.

<u>6.4 Stability.</u>   In practical applications there is one
feature that always needs to be present whenever one is
reconstructing a signal from given data, namely that of stability. A reconstruction method is said to be <u>stable</u> if
small errors in the sample values load to small errors in the
reconstructed signal. The formula at the end of 5.4 shows
that this property occurs for the sampling theorem when all
the sample points are available. Further details are given
in the chapter "An overview of time and frequency limiting".

REFERENCES

1. W. Beckner, Inequalities in Fourier analysis, <u>Ann. of</u>
    <u>Math.</u> 102: 159-182 (1975).

2. J.J. Benedetto, An inequality associated with the uncer-
    tainty principle (submitted).

3. A.L. Cauchy, Mémoire sur diverses formules d'analyse,
    <u>C.R.</u> <u>Acad.</u> <u>Sci.</u> <u>Paris.</u> 12: 283-298 (1841).

4. M.G. Cowling and J.F. Price, Bandwidth versus time-
    concentration: the Heisenberg-Pauli-Weyl inequality,
    <u>SIAM</u> <u>J.</u> <u>Math.</u> <u>Anal.</u> 15: 151-165 (1984).

5. M.G. Cowling and J.F. Price, Generalisations of
    Heisenberg's inequality, <u>in</u> "Harmonic Analysis; Proceed-
    ings 1982", G. Mauceri, F. Ricci and G. Weiss, eds, Lec-
    ture Notes in Mathematics 992, Springer-Verlag, Berlin
    (1983).

6. W.G. Faris, Inequalities and uncertainty principles, <u>J.</u>
    <u>Math.</u> <u>Phys.</u> 19: 461-466 (1978).

7. W.H.J. Fuchs, On the eigenvalues of an integral equation arising in the theory of band-limited signals, J. Math. Anal. Appl. 9: 317-330 (1964).

8. D. Gabor, Theory of communication, J. Inst. Electr. Engnrs. 93(3): 429-457 (1946).

9. G.H. Hardy, A theorem concerning Fourier transforms, J. London Math. Soc. 8: 227-231 (1933).

10. I.I. Hirschman Jr., A note on entropy, Amer. J. Math. 79: 152-156 (1957).

11. A.J. Jerri, The Shannon sampling theorem - its various extensions and applications, Proc. IEEE 65: 1565-1596 (1977).

12. I. Kluvanek, Sampling theorem in abstract harmonic analysis, Mat.-Fyz. Casopsis Sloven. Akad. Vied. 15: 43-48 (1965).

13. V.A. Kotel'nikov, On the transmission capacity of "ether" and wire in electrocommunications, Izd. Red. Upr. Svyazi RKKA (Moscow), 1933.

14. H.J. Landau, H.O. Pollak and D. Slepian, Prolate spheroidal wave functions: Fourier analysis and uncertainty, Bell System Tech. J. : I, 40: 43-64(1961); II, 40: 65-84(1961);   III, 41: 1295-133(1962); IV, 43: 3009-3057(1964); V, 57: 1371-1430(1978).

15. H. Nyquist, Certain topics in telegraph transmission theory, AIEE Trans. 47: 617-644 (1928).

16. J. Pearl, Time, frequency, sequency, and their uncertainty relations, IEEE Trans. Inform. Theory 19: 225-229 (1973).

17. D.P. Petersen and D. Middleton, Sampling and reconstruction of wave-number-limited functions in N-dimensional Euclidean spaces, Inform. and Control 5: 279-323 (1962).

18. J.F. Price, Inequalities and local uncertainty principles, J. Math. Phys. 24: 1711-1714 (1983).

19.   J.F. Price, Minimum conditions for the sampling theorem
      (submitted).

20.   J.F. Price and P.C. Racki, Local uncertainty inequali-
      ties for Fourier series, Proc. Amer. Math. Soc. (to
      appear).

21.   C.E. Shannon, Communication in the presence of noise,
      Proc. IRE 37: 10-21 (1949).

22.   C.E. Shannon and W. Weaver, "The Mathematical Theory of
      Communication," University of Illinois Press, Urbana
      (1949).

23.   D. Slepian, On bandwidth, Proc. IEEE 64: 292-300 (1976).

24.   G.W. Stewart, Problems suggested by an uncertainty prin-
      ciple in acoustics, J. Acoustical Soc. Amer. 2: 325-329
      (1931).

25.   E.T. Whittaker, On the functions which are represented
      by the expansions of the interpolatory theory. Proc.
      Roy. Soc. Edinburgh 35: 181-194 (1915).

26.   J.F. Price, Sharp local uncertainty conditions (submit-
      ted).

FOURIER TECHNIQUES IN TWO DIMENSIONS

R.N. Bracewell

Department of Electrical Engineering
Stanford University
Stanford, California, U.S.A.

FOURIER TRANSFORMS AND IMPULSES IN TWO DIMENSIONS

Consider an area density distribution $f(x,y)$ over a plane expressed as

$$f(x,y) = \delta(r - R).$$

From previous experience telling us that an impulse is located where the argument of the function is zero we expect the density to be zero everywhere except on the circle $r = R$ and we would know everything that

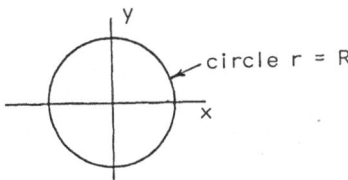

is to be known about $\delta(r-R)$ if we knew either the linear density or the total mass. It helps to grasp the nature of the entities involved if we mention the following consistent set of units.

Quantity	Unit
area density f(x,y)	kg m^{-2}
linear density w(s)	kg m^{-1}
mass $\iint f(x,y)dxdy$	kg
mass $\int w(s)ds$	kg

In this table s is arc length measured along a curve, in this case along the circle of radius R centered on the origin, and w(s) is the linear density or strength at any point on the curve.

To discuss $\delta(r-R)$ we fall back on the rules (see page 70 of my book "The Fourier Transform and its Applications") for interpreting expressions containing $\delta(.)$, which are

(i)   Replace $\delta(.)$ by $\tau^{-1}\Pi(./\tau)$, where Π is the function which is 1 on $(-1/2,1/2)$ and 0 elsewhere.

(ii)   Carry out the operation indicated.

(iii) Proceed to the limit as $\tau \to 0$.

On making the replacement

$$\delta(r - R) \quad \text{becomes} \quad \tau^{-1}\Pi[r - R)/\tau].$$

We can understand this nonimpulsive expression as representing a circular flat-topped wall of thickness τ and height τ^{-1} centered on r = R.   In other words the expression is equal to τ^{-1} between the two concentric circles $(r-R)/\tau = \pm\ 1/2$ and zero outside.   The operation to be carried out is, let us say, to determine the mass expressed by

$$\iint \delta(r - R)dxdy.$$

In accordance with the second rule we evaluate

$$\iint \tau^{-1}\Pi[(r - R)/\tau]dxdy.$$

Changing to polar coordinates we have

$$\int_0^{2\pi} \int_0^\infty \tau^{-1}\Pi[(r - R)/\tau]r\,dr\,d\theta = 2\pi\int_0^\infty \tau^{-1}\Pi[(r - R)/\tau]r\,dr = 2\pi R.$$

One could of course have evaluated this integral geometrically, since it is just the volume of the circular wall, in the form

$$\tau^{-1}[\pi(r + \tau/2)^2 - \pi(R - \tau/2)^2]$$

and have obtained the same answer $2\pi R$. In this example, when the operation of integration has been carried out, the result turns out not to depend on τ any more; therefore taking the limit as $\tau \to 0$ in accordance with rule (iii) makes no further difference. The mass of the ring impulse $\delta(r-R)$ is thus $2\pi R$. From the mass we can get the linear density by dividing the mass by the length of the perimeter. The result is that $\delta(r-R)$ is seen to have unit linear density.

Impulse function of f(x,y)

The most general curvilinear line impulse in the (x,y)-plane is represented as $\delta[f(x,y)]$. From the preceding special cases it is apparent that the line $f(x,y) = 0$ is the locus of the line impulse. Of course $f(x,y) = 0$ is the general equation of a curve in the (x,y)-plane and may very well possess extreme features such as cusps or points of self-intersection or other more or less drastic singularities. But, having decided where the line impulse is located, for the moment we are occupied only with the elementary question of finding out the strength of the line impulse at each point along its length. We have seen that the strength at any point on $f(x,y) = 0$ is the reciprocal of the absolute slope of $f(x,y)$ at that point. The direction of maximum slope is normal to the direction of the null contour. If $f(x,y)$ is given analytically we can calculate the two slope components $\partial f(x,y)/\partial x$ and $\partial f(x,y)/\partial y$ and combine them to obtain $\partial f/\partial n$, the slope in the direction of the normal to the contour.

$$\partial f/\partial n = [(\partial f/\partial x)^2 + (\partial f/\partial y)^2]^{1/2}.$$

Then

$$strength = |\partial f/\partial n|^{-1}_{f=0}.$$

As an example consider

$$f(x,y) = [x^2/a^2 + y^2/b^2]^{1/2} - 1.$$

Viewed as a topographical feature this would be a conical crater with elliptical contours. The zero-height contour $f(x,y) = 0$ would be the central ellipse with semiaxes equal to a and b. It is on this ellipse that we expect to find the

line impulse $\delta\{[x^2/a^2 + y^2/b^2]^{1/2} - 1\}$. Now

$$\partial f/\partial x = x/a^2(f+1)^{1/2}$$

and

$$\partial f/\partial y = y/b^2(f+1)^{1/2}.$$

As we are only interested in slopes at the null contour $f = 0$ we can write

$$(\partial f/\partial x)_{f=0} = x/a^2$$

$$(\partial f/\partial y)_{f=0} = y/b^2.$$

Then $$|\partial f/\partial n|^{-1} = (x^2/a^4 + y^2/b^4)^{-1/2}$$

Fig. 1 shows a special case where $a = 2$ and $b = 1$. A little thought makes it clear why the strength of the elliptical line impulse is twice as great at the ends of the major axis. The slope at $x = a$, $y = 0$ is a^{-1} while the slope at $x = 0$, $y = b$ is b^{-1}. Since the strength of the impulse is the reciprocal of the slope, the strengths at the ends of the major and minor axes are a and b respectively.

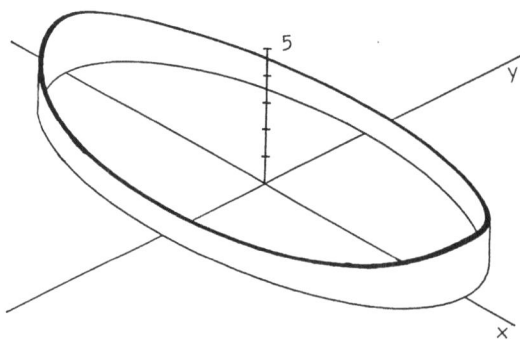

Fig. 1   The elliptical line impulse $\delta[(x^2/a^2 + y^2/b^2)^{1/2} - 1]$ is not of uniform strength.

Two-Dimensional Fourier Transform

In one dimension we define the Fourier transform F(s) of a given function f(x) by

$$F(s) = \int_{-\infty}^{\infty} f(x)e^{-i2\pi sx}dx.$$

We may think of the right-hand side as specifying an operation of analysis as follows. Multiply the given function f(x) by the factor exp(-i2πsx), a function of x containing a parameter s, which is to remain fixed at some constant value for the time being; then integrate the produce over all x. The result is no longer a function of x but does depend on the chosen value of the parameter s. The F value so obtained may now be supplemented by other F values corresponding to other choices of s by repeating the multiplication and integration. The computation performed in this way may be thought of as analysis of f(x) into exponential components. The underlying idea is that exponential components of f(x) not having the chosen s value will contribute products that integrate to nothing whereas the component at the chosen value of s will be analyzed out because the integral of its product can be nonzero.

The inverse relationship

$$f(x) = \int_{-\infty}^{\infty} F(s)e^{i2\pi xs} ds$$

can be thought of as specifying an operation of synthesis in which exponential functions of x with different s values, each having its appropriate amplitude F(s)ds, are summed. The function f(x) is thus synthesized from exponentials of all frequencies s. Of course, we are talking about exponential functions of imaginary argument, a convenient way of handling sinusoids and cosinusoids simultaneously.

In two dimensions, similar viewpoints obtain. The two-dimensional Fourier transform F(u,v) of the two-dimensional function f(x,y) is defined by

$$F(u,v) = \int_{-\infty}^{\infty} \int_{-\infty}^{\infty} f(x,y)e^{-i2\pi(ux+vy)}dxdy.$$

We may view this as an operation of analysis and ask what are we analyzing f(x,y) into. The answer is functions of the form exp[-i2π(ux+vy)] with various amplitudes depending on

the choice of u and v. Splitting the exponential into terms
of the form

$$\cos[2\pi(ux+vy)] \qquad \text{and} \qquad \sin[2\pi(ux+vy)] \qquad (1)$$

or cos $2\pi ux$ cos $2\pi vy$, sin $2\pi ux$ sin $2\pi vy$, sin $2\pi ux$ cos $2\pi vy$
and cos $2\pi ux$ sin $2\pi vy$, we see that functions $f(x,y)$ that are
symmetrical about both the x- and y-axes may be analyzed into
functions of the form cos $2\pi ux$ cos $2\pi vy$. Fig. 2 shows the
character of such a cosine-product function; it is the same
function that describes standing waves arising by perfect
reflection from a rectangular boundary.

The quantity u is the number of waves per unit length in
the x-direction and the quantity v is the number of waves per
unit length in the y-direction. To perform the analysis of a
doubly symmetrical function $f(x,y)$ into such cosine-product
functions, one multiplies $f(x,y)$ by cos $2\pi ux$ cos $2\pi vy$, where
u and v are fixed at certain constant values for the time
being, and integrates the product over the whole (x,y)-plane.
The result is a function only of the chosen u and v and is
the amplitude of the chosen component.

From the synthesis standpoint we may say that a doubly
symmetrical function $f(x,y)$ may be synthesized by the superposition, with appropriate amplitudes, of two-dimensional
cosine-product functions chosen from the full doubly-infinite
range of u and v. The two equations summarizing the above
statements are

$$F(u,v) = \int_{-\infty}^{\infty} \int_{-\infty}^{\infty} f(x,y)\cos 2\pi ux \cos 2\pi vy \, dxdy$$

$$f(x,y) = \int_{-\infty}^{\infty} \int_{-\infty}^{\infty} F(u,v)\cos 2\pi xu \cos 2\pi yv \, dudv.$$

In general, for nonsymmetrical functions, we would need
additional terms as expressed compactly in the two standard
relations

$$F(u,v) = \int_{-\infty}^{\infty} \int_{-\infty}^{\infty} f(x,y)e^{-i2\pi(ux+vy)}dxdy$$

$$f(x,y) = \int_{-\infty}^{\infty} \int_{-\infty}^{\infty} F(u,v)e^{i2\pi(xu+yv)}dudv.$$

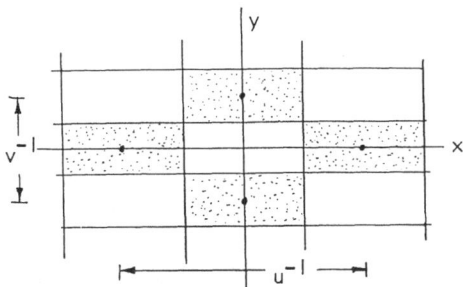

Fig. 2   The "quilted" surface $(\cos 2\pi\ ux)(xos\ 2\pi\ vy)$.
The shaded regions are negative.

The presence of the factor i in front of the term
$\sin 2\pi(ux+vy)$ is worthy of comment. Why should a <u>real</u> function f(x,y) that is not strictly symmetrical, and therefore
requires sine components, require imaginary quantities of the
sine components? If the function is real, the sine component
will have to be real.   This means that the coefficient
F(u,v) cannot be pure real; it will need to be complex in
order to cancel the i that automatically arises from the use
of the exponential notation.

We might describe the pattern illustrated as a "quilted"
surface because it resembles the surface of a quilt whose
lines of stitching are perpendicular to each other.   The
Fourier synthesis theorem is like saying that any surface can
be built up by superposition of quilted surfaces, all
oriented the same way, but having all possible wavelengths in
the two perpendicular directions, and all possible shifts,
with appropriate amplitude.

There is another way of viewing this however. Let us
define a "corrugation" (Figure 3) as a surface generated as
the locus of a straight line that passes through a sinusoid
perpendicular to the plane of the sinusoid. Then the two-
dimensional Fourier synthesis can also be described in terms
of the superposition of corrugations having all possible
wavelengths q^{-1} and all possible orientation ϕ, with
appropriate amplitudes.   The interrelation is

$$q = (u^2+v^2)^{1/2}$$

$$\tan\phi = v/u$$

where q is the spatial frequency of a corrugation (the number of waves per unit length in the direction normal to the wave crests) and ϕ is the angle between that direction and the x-axis. This result follows from substituting.

The corrugation viewpoint provides our clearest way of understanding and remembering the basic significance of the variables u and v. Imagine a sandy desert where the sand dunes have a sinusoidal cross-section and all run parallel to each other. Then if you ride a camel (Figure 4) from west to east, u is the number of crests per unit horizontal distance traveled. The west-to-west distance going west to east is u^{-1}. Likewise if you ride north, the number of crests per meter is v (assuming that distance is measured in meters) and v^{-1} is the number of meters per crest. The true crest-to-crest distance measured perpendicular to the direction of the crest lines is shorter than either u^{-1} or v^{-1}, in general, and of course its the hardest direction in which to ride. The spatial frequency in that direction is greater than in any other direction and may be calculated by vectorial addition of the spatial frequency components along any pair of orthogonal axes.

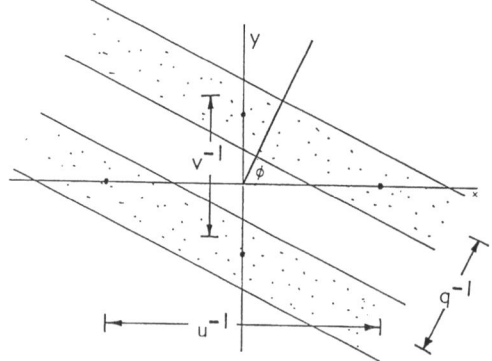

Fig. 3   A "corrugation" $\cos[2\pi(ux+vy)]$. The shaded zones are negative. The spatial frequency in the x-direction is u (in the y-direction ,v) and "the" spatial frequency is q.

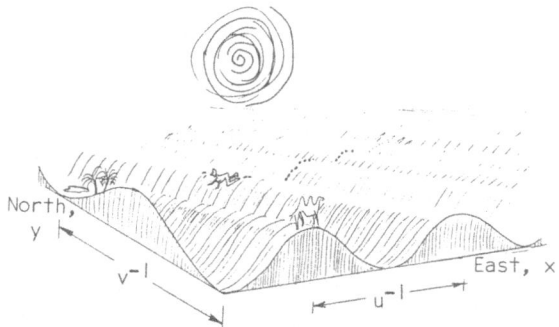

Fig. 4. The x-component of spatial frequency is u,
measured in cycles per unit of x. The spatial period,
or crest-to-crest distance traveling east is u^{-1}
(see camel). The spatial frequency q is measured in the
direction of hardest going (see man).

Since any surface can be analyzed into corrugations of
appropriate amplitude, direction and spatial frequency, it
follows that a quilted component can be so analyzed. We find
that it is just the sum of two corrugations that are equally
inclined to the coordinate axes. To prove this we reflect
the corrugation $\cos[2\pi(ux+vy)]$ in the y-axis to obtain
$\cos[2\pi(ux-vy)]$ and add the two together. Then

$$\cos[2\pi(ux+vy)] + \cos[2\pi(ux-vy)]$$

$$= \cos(2\pi ux)\cos(2\pi vy) - \sin(2\pi ux)\sin(2\pi vy)$$

$$+ \cos(2\pi ux)\cos(\pi vy) + \sin(2\pi ux)\sin(2\pi vy)$$

$$= 2\cos(2\pi ux)\cos(2\pi vy).$$

We see that the two corrugations into which a unit quilt pattern resolves must be of strength 0.5 each.

Figure 5 shows the two corrugations mentioned above
together with parts of the four null lines bounding the central maximum of their combination. By fixing attention on a
point on one of these null lines one can see how the contributions from the two corrugations do indeed cancel along
noninclined loci.

Here is a physical interpretation of the intimate relation between the quilt and corrugation patterns. Suppose
that the x-axis is a perfectly reflecting barrier to waves
that are incident from the NNE and after reflection will be

traveling toward the NNW. The previous figure catches these
waves at the moment when a crest of the incident wavetrain is
at the origin and the same is true for the reflected wave.
Consequently the superposition of the two waves has its maximum at the origin and the full pattern representing the
interference of the two oblique wave trains is the quilt pattern. Of course, the lower half of the figure is to be
ignored. The two traveling waves interfere to set up a
standing wave in the y-direction, for example there will be a
nodal lines of zero disturbance running parallel to the x-
axis, the first one being at a distance $u^{-1}/4$ from the
reflecting barrier on the x-axis. The standing wave is a
consequence of the fact that energy flow to the south is
blocked; on the other hand, there is no barrier to energy
flow to the west, therefore the whole built pattern is to be
regarded as moving to the west. Only at the moment previously frozen in time would the positive maximum be over the
origin. After the time taken for the incident wavetrain to
deliver its next crest to the origin, the quilt pattern will
have moved west by a distance u^{-1}. Since this distance, the
east-west crest-to-crest distance is greater than the
wavelength q^{-1} we see that the quilt pattern moves faster
than the incident waves. This phenomenon can be confirmed
occasionally on beaches where the waves are obliquely
incident (usually they are not) and the situation also arises
in other cases, for example in glass where a ray of light is
reflected when it impinges on the glass-to-air boundary. The
electromagnetic disturbance then propagates along the glass
boundary faster than light. The same happens with microwaves
in a waveguide for the same reason.

Example of transform pairs

    A stock-in-trade of two-dimensional Fourier transform
pairs is useful for illustrating the analysis and synthesis
discussed above and also for illustrating the theorems that
follow.

    Example 1.   As a first example consider the pair

$$^2\delta(x,y) \supset 1$$

which is illustrated in Fig. 6. Thus, a centrally situated
unit two-dimensional impulse transforms into a function that
is equal to unity, independent of u and v. To establish this

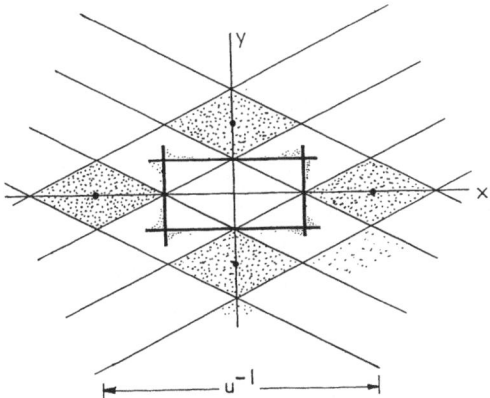

Fig. 5 Two equally inclined corrugations are shown
superimposed. The shaded regions are the intersections
of the negative regions of each corrugation considered
separately but give only an indication of the whereabouts
and shape of the combination. The central positive
bulge of the quilt pattern is indicated in heavy outline.

Fig. 6 The unit two-dimensional impulse and its two-
dimensional Fourier transform. (The small ticks in this
and other illustrations mark unit values along the axes.)

we use the sifting property

$$\int_{-\infty}^{\infty} \int_{-\infty}^{\infty} {}^2\delta(x,y) \ f(x,y)dx \ dy = f(0,0),$$

from which it follows that

$$\int_{-\infty}^{\infty} \int_{-\infty}^{\infty} e^{i2\pi(ux+vy)} du \ dv = {}^2\delta(x,y).$$

To prove this relationship we could if we wished take the
position that the direct transform has already been established and that since the impulse notation must conform with
ordinary notation the inverse transform holds by necessity.
On the other hand one could ask for more direct insight. In

that case the procedure is to consider a sequence of functions, such as for example the sequence generated by the
expression $\exp[-\pi\tau^2(x^2+y^2)]$ as $\tau \to 0$, which has the property
of approaching unity. Each member of the sequence has a cal-
culable transform. We then look at the sequence of
transforms to see whether it is a suitable defining sequence
for $^2\delta(u,v)$. We will be able to do this below.

As to physical interpretation one could say that sound
escaping through a pinhole in a rigid wall spreads uniformly
in all directions. This statement will be made rigorous
later; meanwhile, it should be noted that further interpretation is needed, because the directions available to the sound
reach only to $90°$ from the normal.

The inverse transform can be illustrated in the same
field of acoustics where it means that a wavefront of constant amplitude and phase over some plane radiates only in
the single direction normal to the plane.

This sort of physical interpretation, whether in terms
of acoustics, light, electromagnetic waves or other waves, is
of course of great assistance in thinking about the mathematical material.

Example 2. Taking two half-strength impulses at (a,0)
and (-a,0) and integrating by means of the sifting property
as before we obtain (Fig. 7)

$$0.5\ ^2\delta(x\ a,0) + 0.5\ ^2\delta(x-a,0) \supset \cos 2\pi au.$$

Although the direct transform

$$\int_{-\infty}^{\infty} \int_{-\infty}^{\infty} [0.5\ ^2\delta(x+a,0) + 0.5\ ^2\delta(x-a,0)]e^{-i2\pi(ux+vy)}dx\ dy$$

is very easy to evaluate by splitting into two integrals and
using the sifting property, the inverse transform

$$\int_{-\infty}^{\infty} \int_{-\infty}^{\infty} \cos 2\pi au\ e^{i2\pi(ux+vy)}du\ dv$$

is, as in the previous example, much more complicated.

One learns that it very often happens that the forward
transform and the inverse transform are of unequal difficulty
and that it pays to look at both ways of proceeding before
forging ahead with what may prove to be the hard way.

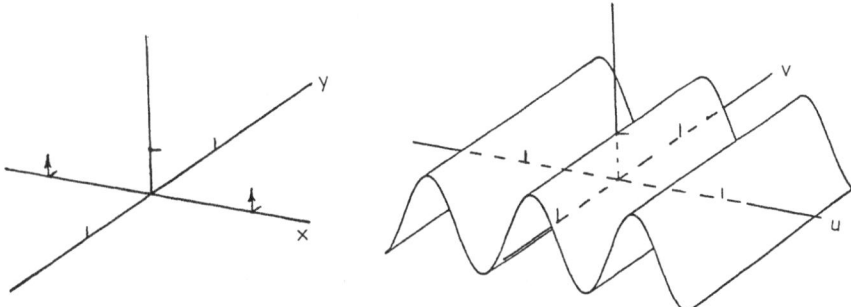

Fig. 7 Symmetrical impulse pair transforms into a cosinusoidal corrugation.

As to physical interpretation, here we are dealing with the cosinusoidal interference pattern produced when waves escape through two pinholes, or from two point sources, with the same phase and amplitude at each source point. As before, the inverse transform also has an interpretation and one that is different in character from the interpretation of the direct transform. If we could impose a cosinusoidal amplitude variation, of spatial period a^{-1}, over a plane wavefront then, according to the transform pair, radiation would be launched in just two directions, equally but oppositely inclined to the normal to the plane. The angle of launch, which is fixed by the spatial period A^{-1}, can easily be deduced.

Example 3.   A two dimensional Gaussian function

$$\exp[-\pi(x^2+y^2)],$$

which may also be written $\exp(-\pi r^2)$ in terms of the polar coordinate r, arises in innumerable connections. It is its own Fourier transform (Fig. 8),

$$e^{-\pi r^2} \supset e^{-\pi q^2}.$$

In this statement we use q as the radial polar coordinate in the (u,v) - plane; thus

$$r^2 = x^2 + y^2 \quad \text{and} \quad q^2 = u^2 + v^2.$$

The coefficient π is included in the position shown partly to

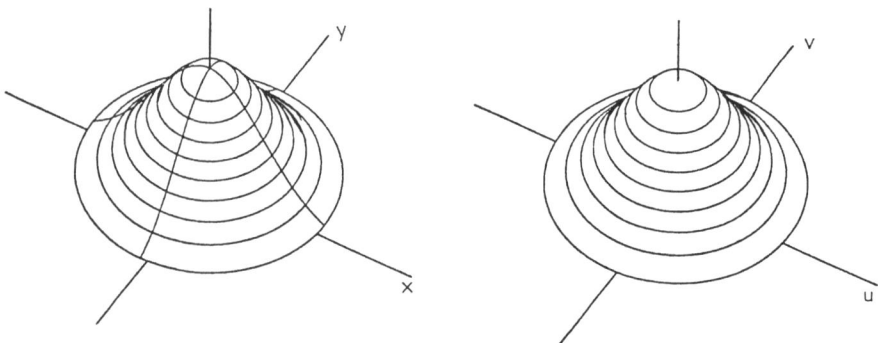

Fig. 8 The unit Gaussian hump is its own Fourier
transform.

have the convenience of symmetry, which makes it easier to
remember this pair, and partly because $\exp(-\pi x^2)$ has unit
area and $\exp(-\pi r^2)$ has unit volume.

To derive the result requires evaluating the integral in
a conventional manner or, as is customary with standard
integrals, looking it up in a table of integrals.

Example 4.   A unit rectangle function $^2\Pi(x,y)$
transforms into the product of two sinc functions as follows
(Figure 9)

$$^2\Pi(x,y) \supset \text{sinc } u \text{ sinc } v.$$

Since in one dimension $\Pi(.)$ transforms into sinc $(.)$ one
might suspect that there is a very simple direct derivation
of this example. It will be given below. Radiation from a
square aperture has a directional dependence connected with
sinc u sinc v. As we are often concerned with power rather
than with amplitude in radiation problems, it will often be
$\text{sinc}^2 u \text{ sinc}^2 v$ that is encountered. Apart from any importance
that rectangular structures may have in the world of
engineering, the two-dimensional rectangle function, which in
speech is more conveniently called rect x rect y, arises in
various other ways. For example, it is a two-dimensional
gate function which, by multiplication selects out a portion
of a field for retention while putting the surrounding function values to zero.

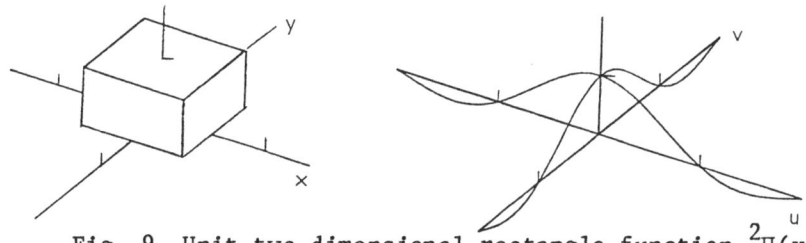

Fig. 9 Unit two dimensional rectangle function $^2\Pi(x,y)$
transforms into a function sinc u sinc v that is
suggested by its two principal cross-sections.

Example 5.   Corresponding to the square aperture is
the equally important circular aperture. Interpreted in two
dimensions, the unit rectangle function of radius, rect r or
$\Pi(r)$, represents a function that is equal to unity over a
central circle of unit diameter and zero elsewhere. Just as
the Fourier transform of a rectangle function in one dimension is a sinc function, so the two-dimensional transform of
rect r is a jinc function. Naturally, a function as fundamental as the diffraction pattern of a circular aperture is
itself intrinsically simple in nature. However, the understanding of the jinc function will be deferred for the
moment. Here we content ourselves with introducing the formal definition jinc q $\Delta J_1(\pi q)/2q$, where J_1 is the Bessel
function of the first kind of order unity. Then (Figure 10)

$$\text{rect } r \supset \text{jinc } q.$$

Example 6.   To show what happens when the function
$f(x,y)$ is independent of y, i.e. when its representation
would be a cylinder (in the sense of a surface generated by a
straight line moving parallel to itself and passing through a
fixed curve), consider $\exp(-\pi x^2)$. Its two-dimensional
transform is representable by a nonuniform line impulse running along the u-axis. In Figure 11 this is represented by a
blade of height equal to the strength of the line impulse at
each value of u. The Fourier transform pair is (Figure 11)

$$\exp(-\pi x^2) \supset \exp(-\pi u^2)\delta(v).$$

Note that this is a case where the sign   is being used for
"has two-dimensional Fourier transform" and that a possibility of confusion might exist with the other meaning "has
Fourier transform". To avoid such risk, where it exists, one
can write $^2\supset$.

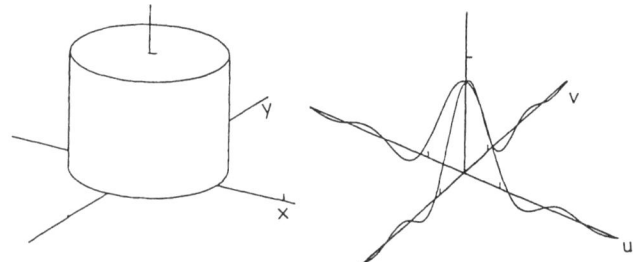

Fig. 10   Introducing the jinc function, the circular
analogue of the sinc function and two dimensional
Fourier transform of the unit pillbox function $\Pi(r)$.

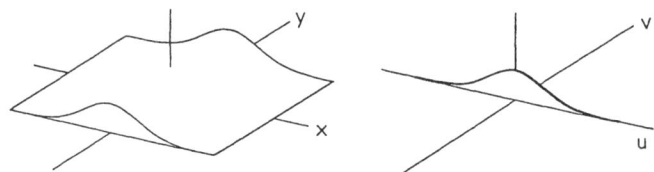

Fig. 11   The Gaussian cylinder $\exp(-\pi x^2)$
transforms into a Gaussian blade.

Example 7.   Finally we have a pair that could be generated from the preceding one by considering a sequence of
Gaussian cylinders $\tau^{-2} \exp(-\pi x^2/\tau^2)$ which, as $\tau \to 0$, would
be a suitable defining sequence for $\delta(x)$. If we regard $\delta(x)$
as a function of two dimensions, but independent of y, we are
talking about a unit line impulse running along the y-axis.
Its Fourier transform is $\delta(v)$, a unit line impulse running
along the u-axis. Thus (Figure 12)

$$\delta(x) \; {}^2\!\!\supset \delta(v).$$

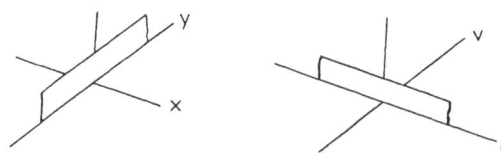

Fig. 12   A unit line impulse on the y-axis transforms
into a unit line impulse on the u-axis.

AUTOCORRELATION IN TWO DIMENSIONS

From the expression

$$\int_{-\infty}^{\infty} \int_{-\infty}^{\infty} f(x´,y´)f(x+x´, y+y´)dx´dy´$$

we see that we start with a given function $f(x,y)$. We rela-
bel the axes $x´$ and $y´$ because we are going to integrate over
the plane, whereupon the dummy variables $x´$ and $y´$ will
disappear. We wish to be left with a function of x and y.
The given function now appears as $f(x´,y´)$. Now the second
factor $f(x+x´, y+y´)$ is a replica of $f(x´,y´)$ but displaced
by an amount x to the left and an amount y downwards. We may
represent this situation by copying the picture $f(x´,y´)$ on a
piece of tracing paper and displacing it as described. If we
push a pin through the displaced copy and into the original
we have the value $f(x+x´, y+y´)$ at one pinhole and the value
$f(x´,y´)$ at the other. These values are to be multiplied
together and we can imagine the product as a new function
covering the $(x´,y´)$-plane. According to the defining
expression for the autocorrelation we are now to perform the
double integral or find the volume under the product function. That is a lot of work involving multiplications all
over the plane followed by summing of products all over the
plane. Even so, the result is merely a single value of the
autocorrelation, namely for the particular (x,y) describing
the displacement. To get another value we have to displace
the function to a new position, multiply throughout and find
the volume under the product again. To obtain the whole
autocorrelation function we have to repeat over and over
until all desired values of x and y are covered.

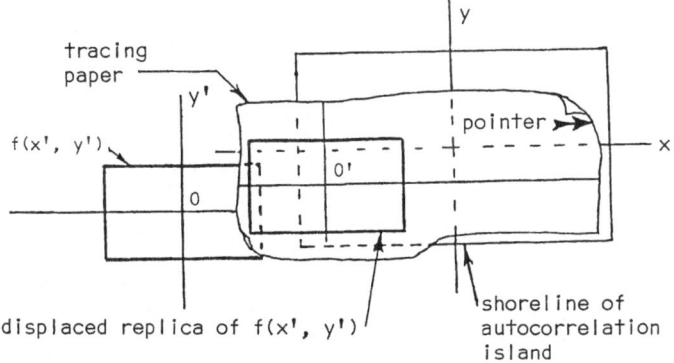

Fig. 13 Tracing paper construction for autocorrelation

Figure 13 shows an original function and its displaced replica. They overlap, for the chosen displacement 00´, only over a thin rectangle. The product is to be integrated over this region of overlap and recorded. Then a new displacement is chosen. A convenient way to record the integral is to make a dot on the (x,y)-plane in a location indicated by a pointer carried by the tracing paper and to write the value alongside.

It is a very good idea to get a piece of tracing paper and experiment with this mechanical construction because it develops certain intuition that remains useful later. A number of general points can be made. For example, if the original function f is zero outside a certain boundary, then f ** f will also be zero outside some other boundary. The shoreline of this autocorrelation island may be explored rapidly by moving the replica so as to keep it tangent to the original while simultaneously drawing the locus traced by the pointer. This can be done by making a reinforced hole in the paper at the tip of the pointer and putting the point of a pencil through as illustrated in Figure 14.

It is quite possible to dispense with the use of tracing paper and visualize the motion of an imaginary pencil point as it traces out the boundary of the autocorrelation island. However, to gain this ability it is necessary even for people with good powers of visualization to practise with tracing

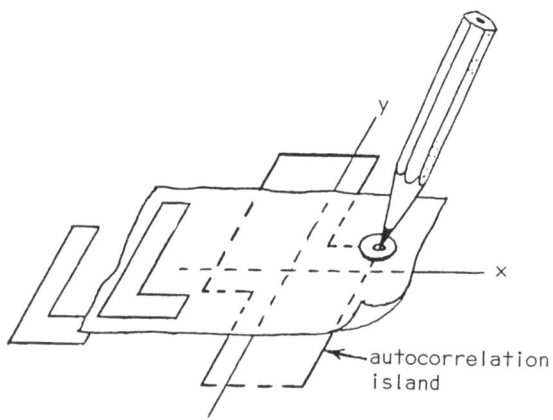

Fig. 14   Method of tracing shoreline of autocorrelation island.

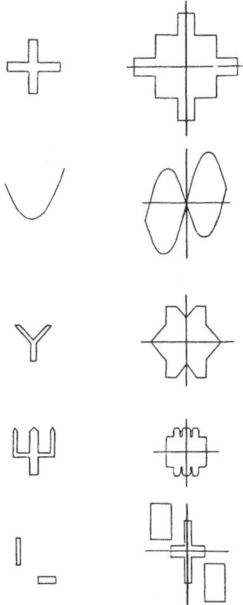

Fig. 15 Various outlines (left) and their
autocorrelation islands (right). An error
has been introduced to provide an exercise.

paper. Figure 15, which exhibits a number of examples, will
enable the reader to determine his current level of ability.
For practise, trace the diagrams on the left, verify the
autocorrelation islands on the right and discover which one
contains an error which has been incorporated to test your
ability to dispense with the tracing paper.

A feature noticeable from the selection of autocorrelation islands is a certain symmetry with respect to the axis
that is perpendicular to the x- and y-axes and passes through
the origin. If any of the autocorrelation islands is rotated
about the perpendicular axis the shape repeats each half
turn. The islands are said to possess a two-fold axis of
symmetry. Furthermore, a little thought shows that the autocorrelation function values themselves have the same symmetry; that is, the value of the autocorrelation at any point
(x,y), not necessarily, on the boundary, is the same as the

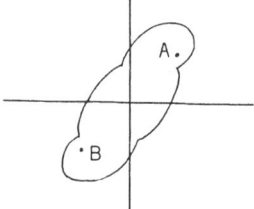

Fig. 16 Under two-fold rotational symmetry,
and autocorrelation repeats after half a turn
about the origin. Thus the value at A is the
same as at the diametrically opposite point B

value at the diametrically opposite point $(-x, -y)$. A formal
proof can easily be given by referring to the definition
integral, replacing x by $-x$ and y by $-y$ and showing by sub-
stitution of variables that the integral is unchanged.   Thus
$\gamma(-x, -y) = \gamma(x, y)$ as illustrated in Figure 16.

A rotation π about the perpendicular axis can be
achieved in another way.   Take a sheet of paper printed on
one side only.   Mark x- and y- axes.   Now flip the paper
about the x- axis, rotating half a turn about that axis.
Then flip about the y- axis half a turn.   The net result, as
you see, is half a turn about the perpendicular axis.   The
reason is that reversal of the sign of x followed by reversal
of the sign of y takes each point (x,y) to the diametrically
opposite point $(-x,-y)$.   A further way of looking at this is
to understand that it makes no difference whether the replica
is displaced a certain distance to say the south-west or the
same distance to the north-east.   Either way, the configura-
tion of both shapes taken together is the same.

A final general feature of the autocorrelation is
independence of the origin of $f(x,y)$, that is to say if you
move $f(x,y)$ to another place it still has the same autocorrelation.   This is obvious from the tracing paper construction
and easy to prove mathematically by substituting $f(x+a, y+b)$
into the definition integral and showing that the integral is
unchanged.

Lazy pyramid and Chinese hat function

As examples of autocorrelation that may be worked out analytically from the integral definition we take two cases that are frequently met in a variety of circumstances.

Suppose we have a function $f(x,y)$ that could be described as a square-table function of unit height and unit side. To express this function algebraically we could write

$$f(x,y) = \begin{cases} 1 & |x| < 1/2 \text{ and } |y| < 1/2 \\ 0 & \text{otherwise} \end{cases}$$

Admittedly this notation is a little awkward because the inequalities have to be studied rather carefully to see that the function intended is quite simple. Furthermore, the notation is rather lengthy. For these reasons it is convenient to make use of the two-dimensional unit rectangle function $^2\Pi(x,y)$, illustrated in Figure 17, which may be defined by

$$^2\Pi(x,y) = \begin{cases} 1 & \text{inside central unit square} \\ 0 & \text{outside} \end{cases}$$

With this notation agreed upon, suppose we wish to autocorrelate a function $f(x,y)$ such that

$$f(x,y) = {}^2\Pi(x,y).$$

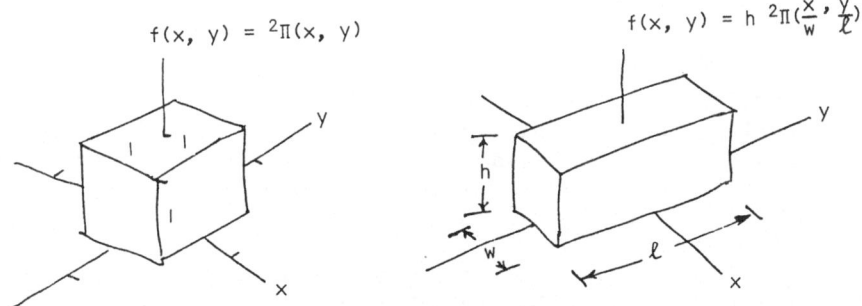

$f(x, y) = {}^2\Pi(x, y)$

$f(x, y) = h\ {}^2\Pi(\frac{x}{w}, \frac{y}{\ell})$

Fig. 17 The two-dimensional unit rectangle function (left) and an example of how the rectangle function notation may be employed to describe a block of width w, length 1 and height h (right).

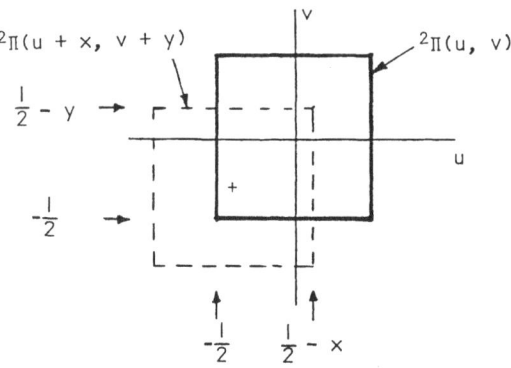

Fig. 18   The autocorrelation of the two-dimensional unit rectangle function is equal to the area of overlap.

From the definition integral for the autocorrelation $c(x,y)$ we have

$$c(x,y) = \int_{-\infty}^{\infty} \int_{-\infty}^{\infty} {}^{2}\Pi(u,v)\, {}^{2}\Pi(u+x,\ v+y)\,du\,dv.$$

Assume for the moment that x and y are both positive. Then

$$c(x,y) = \begin{cases} \int_{-1/2}^{1/2-y}\int_{-1/2}^{1/2-x} du\,dv & |x| \text{ and } |y| < 1/2 \\ 0 & \text{otherwise} \end{cases}$$

$$= \begin{cases} (1-|x|)(1-|y|) & |x| \text{ and } |y| < 1/2 \\ 0 & \text{otherwise} \end{cases}$$

To see how the limits of integration are arrived at is a topological rather than an algebraic exercise and is most easily carried out with a sliding piece of tracing paper. Figure 18 illustrates a fixed situation where x = 4/5 and y = 3/5 and will permit the limits of integration to be verified. Clearly the integral is equal to the area of the rectangle of overlap whose dimensions are 1-x and 1-y; thus we reach the same conclusion as when we use the result $\iint du\,dv = uv$ and substitute limits.

In order to carry out these arguments we assumed that x and y were both positive. If one or both are negative the limits change. For example, if x is negative, the integral with respect to u becomes

$$\int_{-1/2-x}^{1/2} \text{ instead of } \int_{-1/2}^{1/2-x}.$$

There are four cases to consider altogether which is not only
tedious but fraught with risk of algebraic error if all the
substitutions are reasoned out and carried out. Therefore,
we apply more fundamental reasoning to argue that the sign of
x and y should not affect the autocorrelation, which must be
symmetrical with respect to both x and y, and that therefore
x and y may be replaced by $|x|$ and $|y|$ respectively in the
limited expression first obtained. Hence

$$c(x,y) = (1 - |x|)(1-|y|).$$

The resulting function is the lazy pyramid, that is, it is
like a pyramid but one which has slumped along its four slop-
ing edges. The lines of steepest descent along the middle of
each face are, however, straight (Figure 19).

Our second example is the autocorrelation of a circular
pillbox function, one that is equal to unity inside the central circle of unit diameter and zero outside.

There is no need to invent new notation to cover this
case because we are talking about a unit rectangle function
of radius $\Pi(r)$. By definition, this rectangle function is
zero for $r > 1/2$ and equal to unity in the range 0 to 1/2.
Very often there is no need to talk about negative values of
the radial coordinate r but if negative values of r were
introduced the notation would be unaffected. Although a
two-dimensional notation because of the circular symmetry
that makes it unnecessary to mention the azimulthal coordinate θ.

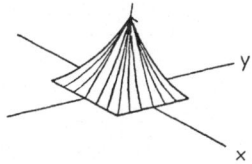

Fig. 19   The "lazy pyramid" $(1-|x|)(1-|y|)$,
the autocorrelation of a unit square-table function
$\Pi(x,y)$.

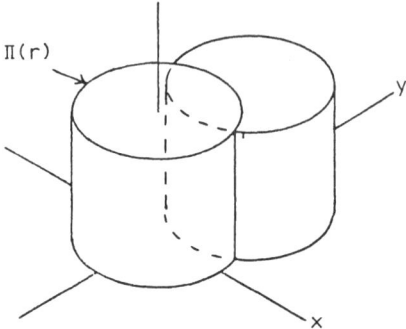

Fig. 20   The autocorrelation of the unit rectangle
function of radius, or unit pillbox, is determined
by integrating the product of $\Pi(r)$ with a shifted
replica, over the region of overlap.

To picture the geometry involved in calculating the
autocorrelation refer to Figure 20.

Both functions (the original and the shifted replica)
being unity over the region of overlap, their product is
unity. Therefore the autocorrelation will just be equal to
the area of overlap which can be derived simply by reference
to Figure 21. The centers of two circles are at C and C´.
The area of overlap may be calculated by noticing that the
shaded area, which is one-quarter of the whole, is expressi-
ble as the difference between the circular sector CPR and the
triangle CPQ.

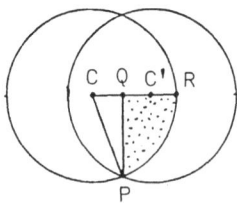

Fig. 21   The area of overlap of the two unit circles
is four times the shaded area PQR. Each circle has
radius 1/2 and the separation CC´ is r.

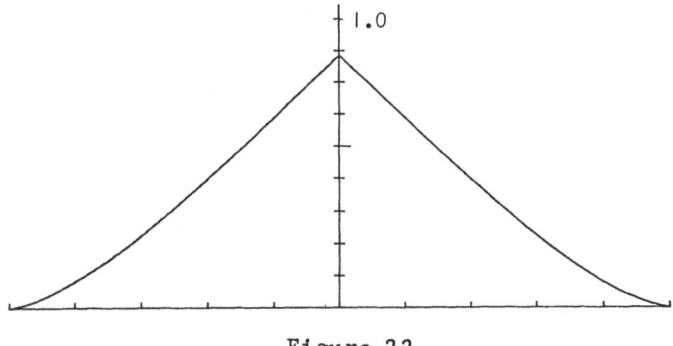

Figure 22

The separation CC' between the centers of the two circles is the value of r to which the calculation will refer.
Thus

chat r = 4(sector CPR - triangle CPQ)

$$= 4\;\frac{1}{2}CP^2\cos^{-1}(CQ/CP) - \frac{1}{2}(CC'/2)\;(\frac{1}{2})^2 -(\frac{1}{2}r)^2]^{1/2}$$

$$= \frac{1}{2}\cos^{-1} - \frac{1}{2}r(1-r^2)^{1/2}.$$

Because of the shape of this function (Figure 22) it is
referred to as the Chinese hat function. Two simple properties are useful for reference:

$$\text{chat } 0 = \pi/4,$$

$$2\pi \int_0^1 r \text{ chat } r \; dr = (\pi/4)^2.$$

The diameter of course is 2, being twice that of the original
circle.

Central value and volume of autocorrelation

Two simple theorems are of considerable value for checking and for normalizing, the first referring to the central
value $c(0,0)$ and the second to the volume under the autocorrelation function $c(x,y)$. By putting $x = y = 0$ in the
definition integral we obtain

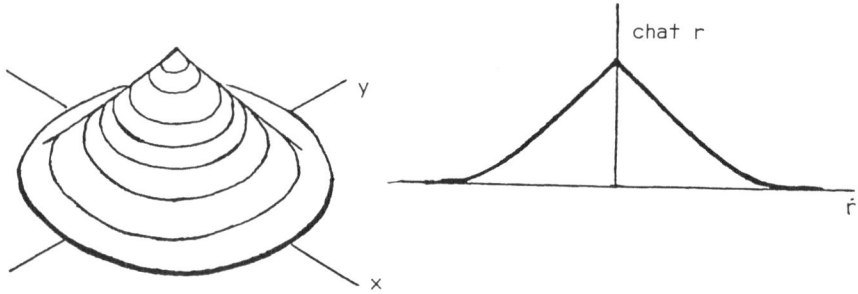

Fig. 23   A pictorial view of the Chinese hat function
(left) and its cross-section (right).

$$c(0,0) = \int_{-\infty}^{\infty} \int_{-\infty}^{\infty} [f(x,y)]^2 dxdy.$$

The theorem for volume is

$$\int_{-\infty}^{\infty} \int_{-\infty}^{\infty} c(x,y)dxdy = [\int_{\infty}^{\infty} \int_{\infty}^{\infty} f(x,y)dxdy]^2$$

Thus the central value of the autocorrelation is the
volume under the square of the original function. The volume
under the autocorrelation is the square of the volume under
the original function. Applying these theorems to the chat
function we immediately verify that chat $0 = \pi/4$ and that the
volume under chat r is $(\pi/4)^2$.

To derive the volume theorem requires the evaluation of

$$\iint c(x,y)dxdy = \iint [\iint f(x´,y´)f(x´+x, y´+y)dx´dy´]dxdy.$$

To perform this forbidding quadruple integration we rearrange
things in the form

$$\iint f(x´,y´)[\iint f(x´+x, y´+y)dxdy]dx´dy´.$$

Since the integral is square brackets is to be performed with
respect to x and y it is permissible to remove the factor
$f(x´,y´)$, which does not depend on x or y, outside the inner
integrals. The mathematical condition for changing the order
of integration in a case such as this is that the integral of
f should exist, which is essentially the only situation we
are interested in. Now the integral in square brackets is

just the volume under f because the volume is invariant under
a shift of original that is to say f(x,y) has the same volume
as the shifted version f(x+x´, y+y´). The square brackets
thus being a constant independent of x´ and y´ may be removed
outside the remaining integral which we see amounts to a
second factor that is also equal to the volume of f. Hence
the volume of the autocorrelation is the square of the volume
of the original function.

The autocorrelation volume theorem is a special case of
a more general theorem that is familiar in its one-
dimensional form as a relation between the area under a convolution and the areas under the two functions entering into
the convolution. Thus, if h(x) = f(x) - g(x), then

$$\int_{-\infty}^{\infty} h(x)dx = \int f(x)dx \int g(x)dx.$$

The area under the convolution h(x) is equal to the product
of the areas under the given functions f(x) and g(x). Likewise for the convolution sum, in the discrete case, the sum
of the convolution string is the product of the sums of the
two factors, a relationship that is useful all the time for
numerical checking when short sequences are being dealt with
by hand. The two-dimensional generalization is

$$\int_{-\infty}^{\infty} \int_{-\infty}^{\infty} f \ast\ast g \ dxdy = \int_{-\infty}^{\infty} \int_{-\infty}^{\infty} f \ dxdy \int_{-\infty}^{\infty} \int_{-\infty}^{\infty} g\,dxdy;$$

the volume under the convolution equals the product of the
separate volumes. As autocorrelation is a special case of
convolution, the autocorrelation volume theorem follows. We
use it continually for checking calculations and for determining normalizing factors.

SOME MATHEMATICAL METHODS FOR SPECTRUM ESTIMATION

John J. Benedetto

Department of Mathematics
University of Maryland
College Park, Maryland, U.S.A.

In these lectures we state the spectrum estimation problem (Section 5) and develop periodogram (Section 6) and maximum entropy (Section 7) methods used for its solution. The
two methods are really general approaches for dealing with
the problem, and most spectrum estimation algorithms fall
into one or the other of these categories. Our presentation
is decidedly mathematical, but our bibliography contains
references for many of the specific engineering and statistical algorithms derived from the techniques outlined herein.

Section 1 sets down the mathematical notation and concepts for our treatment. Sections 2 and 4 outline those
parts of generalized harmonic analysis which we need, and
Section 3 does the same for stochastic processes.

1. MATHEMATICAL SETTING

A function $F: \mathbb{R} \to C$ has <u>bounded variation</u> on $\mathbb{R}$,
written $F \in BV$, if
$$\sup \Sigma \; |F(\gamma_{j+1}) - F(\gamma_j)| < \infty,$$
where the supremum is taken over all finite sets
$\{\gamma_1, \ldots, \gamma_m\} \subseteq \mathbb{R}$. F is <u>locally absolutely continuous</u> on $\mathbb{R}$,
written $F \in AC_{loc}$, if for all $\alpha, \beta, \varepsilon$, there exists $\delta > 0$
such that for all $\{(\lambda_j, \gamma_j) \subseteq [\alpha, \beta]: j = 1, \ldots, n\}$, a disjoint
collection,

73

$$\Sigma \ |\gamma_j - \lambda_j| < \delta \quad \text{implies} \quad \Sigma \ |F(\gamma_j) - F(\lambda_j)| < \epsilon.$$

The <u>Heaviside function</u> $H_\gamma : \mathbb{R} \to \mathbb{C}$ at $\gamma \ \epsilon \ \mathbb{R}$ is

$$H_\gamma(\omega) = \begin{cases} 1, & \omega > \gamma \\ 0, & \omega < \gamma. \end{cases}$$

$X \subseteq \mathbb{R}$ is a set of <u>measure</u> 0 if $\forall \epsilon > 0$, there exists $\{(\alpha_j, \beta_j) : j = 1, \ldots\}$ such that

$$X \subseteq \cup(\alpha_j, \beta_j) \quad \text{and} \quad \Sigma \ (\beta_j - \alpha_j) < \epsilon.$$

A property is valid <u>almost everywhere</u>, written a.e., if it is true for all $\omega \ \epsilon \ \mathbb{R} \setminus X$ where X is a set of measure 0.

<u>Theorem</u> <u>1.1</u>.   Each $F \ \epsilon \ BV$ can be written in the form

$$F = F_{ac} + F_{sc} + \Sigma \ d_\gamma H_\gamma,$$

where $F_{ac} \ \epsilon \ BV \cap AC_{loc}$, $\Sigma \ |d_\gamma| < \infty$, F_{sc} is continuous on $\mathbb{R}$, and $dF_{sc}/d\omega = 0$ a.e.

$C_c^\infty(\mathbb{R})$ is the space of infinitely differentiable functions with compact support. For $F, G \ \epsilon \ C_c^\infty(\mathbb{R})$, integration by parts yields

$$\int F^{(n)} G = (-1)^n \int F G^{(n)}$$

which we write as

$$\langle F^{(n)}, G \rangle = (-1)^n \langle F, G^{(n)} \rangle. \tag{1.1}$$

If $F : C_c^\infty(\mathbb{R}) \to \mathbb{C}$ is a linear functional on $C_c^\infty(\mathbb{R})$, then (1.1) defines the <u>nth distributional derivative</u> $F^{(n)}$ of F. F is a <u>distribution</u> if $\lim \langle F, G_\alpha \rangle = 0$ for every directed system $\{G_\alpha\} \subseteq C_c^\infty(\mathbb{R})$ which is supported by a fixed compact set and which converges uniformly to 0 along with all of its derivatives.

<u>Definition</u> <u>1.1</u>. $M(\mathbb{R}) = \{F^{(1)} : F \ \epsilon \ BV\}$, the set of first <u>distributional</u> derivatives of BV, is the space of <u>bounded Radon measures</u>. $\delta_\gamma \ \epsilon \ M(\mathbb{R})$, defined equivalently by $\langle \delta_\gamma, G \rangle = G(\gamma)$ or $\delta_\gamma = H_\gamma^{(1)}$, is the <u>Dirac</u> δ-measure. If

$F \varepsilon AC_{loc}$ then $F^{(1)}$ is the ordinary pointwise derivative, i.e.,

$$\langle F^{(1)}, G \rangle = \int F'G = -\int FG'.$$

Remark 1.1.   a. If $C_0(\mathbb{R})$ is the space of continuous functions vanishing at $\pm\infty$ and is equipped with the sup-norm $\| \ \|_\infty$ then $M(\mathbb{R}) = C_0(\mathbb{R})'$, the dual of $C_0(\mathbb{R})$. The space of Lebesgue integrable functions can be defined as

$$L^1(\mathbb{R}) = \{F^{(1)}: F \varepsilon BV \cap AC_{loc}\} \subseteq M(\mathbb{R}).$$

b. If $C_b(\mathbb{R})$ is the space of bounded continuous functions equipped with $\| \ \|_\infty$ then $C_b(\mathbb{R})'$ is the space of finitely additive bounded regular measures on the Borel sets $\mathbb{B}(\mathbb{R})$. Also, $L^1(\mathbb{R})' = L^\infty(\mathbb{R})$, the space of essentially bounded measurable functions normed by $\| \ \|_\infty$; and $L^\infty(\mathbb{R})'$ is the space of finitely additive bounded measures on $\mathbb{B}(\mathbb{R})$. Thus, $M(\mathbb{R}) \subseteq C_b(\mathbb{R})'$. Also, $C_0(\mathbb{R}) \subseteq C_b(\mathbb{R})$, but, because the imbedding is not dense, we can not conclude that $C_b(\mathbb{R})' \subseteq C_0(\mathbb{R})'$.

Because of this discussion, Theorem 1.1 can be reformulated as follows.

Theorem 1.2.   a. Each $S \varepsilon M(\mathbb{R})$ can be written in the form

$$S = S_1 + \mu_{sc} + \Sigma \, d_\gamma \delta_\gamma \ ,$$

where $S_1 \varepsilon L^1(\mathbb{R})$, $\Sigma \, |d_\gamma| < \infty$, and μ_{sc} is designated the continuous singular part of S.

b. Using the notation from Theorem 1.1, we have

$$S = F^{(1)}, \ S_1 = F^{(1)}_{ac}, \text{ and } \mu_{sc} = F^{(1)}_{sc}.$$

c. For each $G \varepsilon C_0(\mathbb{R})$ the duality $\langle S, G \rangle$ is denoted by

$$\int G(\omega)dS(\omega) = \int G(\omega)S_1(\omega)d\omega + \int G(\omega)dF_{sc}(\omega) + \Sigma \, d_\gamma G(\gamma),$$

where the integrals on the right-hand side are Lebesgue and Stieltjes, respectively.

Definition 1.2.   If $G \in C_0(\mathbb{R})$ then $(\tau_\gamma G)(\omega) = G(\omega-\gamma)$ and the <u>translation</u> $\tau_\gamma S$ is defined by $\langle \tau_\gamma S, G \rangle = \langle S, \tau_{-\gamma} G \rangle$. The <u>convolution</u> $S_1 * S_2$, for $S_j \in M(\mathbb{R})$, is defined by

$$\langle S_1 * S_2, G \rangle = \langle S_1(\gamma), \langle S_2(\omega), G(\gamma+\omega) \rangle \rangle.$$

Thus, $\delta_\gamma * S = \tau_\gamma S$.

In Theorem 1.2a, $\Sigma \, d \, \delta_\gamma$ is the <u>discrete</u> part of the measure S. Spectrum estimation is concerned with determining the discrete parts of measures.

2. GENERALIZED HARMONIC ANALYSIS

The <u>Fourier</u> <u>transform</u> of $f \in L^1(\mathbb{R})$ is

$$F(\omega) = \int_{-\infty}^{\infty} f(t)e^{-2\pi it\omega}dt = \hat{f}(\omega).$$

A fundamental property of Fourier transforms is

$$(f*g)^\wedge = FG \quad \text{and} \quad (fg)^\wedge = F*G, \tag{2.1}$$

where $f \leftrightarrow F$ and $g \leftrightarrow G$ are Fourier pairs. Also,

$$f(t) = \int_{-\infty}^{\infty} F(\omega)e^{2\pi it\omega}d\omega = F^\vee(t) \tag{2.2}$$

for the Fourier pair $f \leftrightarrow F$; and $(F*G)^\vee = F^\vee G^\vee$. These formulas, as well as some in Example 2.1, are not unconditionally true but are valid for the functions we shall consider. The <u>Fourier</u> <u>transform</u> $\hat{S}$ of $S \in M(\mathbb{R})$ is defined by

$$\langle \hat{S}, g \rangle = \langle S, G \rangle \tag{2.3}$$

for all $g \in C_c^\infty(\mathbb{R})$.

Example 2.1.   a.   For $S \in M(\mathbb{R})$,

$$(\tau_\gamma S)^\wedge(t) = e^{-2\pi i t \gamma} \hat{S}(t), \qquad (2.4)$$

$$(e^{2\pi i u \omega} S(\omega))^\wedge(t) = (\tau_u \hat{S})(t)$$

and

$$\frac{d}{dt}\, \hat{S}(t) = (-2\pi i \omega S(\omega))^\wedge(t). \qquad (2.5)$$

b. $\hat{\delta}_\gamma(t) = e^{-2\pi i t \gamma}$.

c. $H_\gamma \notin M(\mathbb{R})$ but $\hat{H}_\gamma$ can be defined by (2.3).

In fact, writing $H = H_0$ and $\delta = \delta_0$ we have

$$\hat{H}(t) = \frac{1}{2\pi i}\, pv(\frac{1}{t}) + \frac{1}{2}\, \delta.$$

To see this, it suffices to compute

$$\lim_{R\to\infty} \int_{\frac{1}{R}<|t|<R} \frac{e^{2\pi i t \omega}}{t}\, dt = \lim_{R\to\infty} i \int_{-R}^{R} \frac{\sin 2\pi t \omega}{t}\, dt = 2\pi i H_0(\omega) - \pi i,$$

$$\omega \neq 0.$$

d. If $f_r(t) = e^{-\pi r t^2}$, $r > 0$, then

$\hat{f}_r(\omega) = \frac{1}{\sqrt{r}} e^{-\pi\omega^2/r}$; in particular, $\hat{f}_1 = f_1$.

e. If $f_r(t) = e^{-2\pi r|t|}$, $r > 0$, then

$$\hat{f}_r(\omega) = \frac{r}{\pi(r^2+\omega^2)}.$$

$S \varepsilon M(\mathbb{R})$ is a <u>positive measure</u>, written $S \geqslant 0$, if $\langle S,G\rangle \geqslant 0$ for each non-negative $G \varepsilon C_0(\mathbb{R})$. A function $P: \mathbb{R} \to \mathbb{C}$ is <u>positive definite</u>, written $P \gg 0$, if $\Sigma\, c_j \overline{c}_k P(t_j-t_k) \geqslant 0$ for each finite

collection $c_1,\ldots,c_n \varepsilon \mathbb{C}$ and $t_1,\ldots,t_n \varepsilon \mathbb{R}$. If $S \geqslant 0$

then $P = \hat{S} \gg 0$; in fact,

$$\Sigma\ c_j\bar{c}_k P(t_j - t_k) = \int |\Sigma c_j e^{-2\pi i t_j \omega}|^2\ dS(\omega).$$

The converse is Bochner's theorem:

Theorem 2.1.   Let P be continuous on $\mathbb{R}$. $P \gg 0$ if and only if $\exists S > 0$ for which $\hat{S} = P$ on $\mathbb{R}$.

Also, it is not difficult to verify -

Proposition 2.1.   Let P be continuous on $\mathbb{R}$. $P \gg 0$ (and therefore is bounded) if and only if $P \in C_b(\mathbb{R})$ and for all $f \in L^1(\mathbb{R})$,

$$\int\int P(t+u)f(t)\tilde{f}(u)dtdu = \int P(t)f*\tilde{f}(t)dt > 0, \qquad (2.6)$$

where $\tilde{f}(t) = \overline{f(-t)}$.

Given $x \in L^\infty(\mathbb{R})$, its **autocorrelation** is the function

$$P(t) = \lim_{T \to \infty} \frac{1}{2T} \int_{-T}^{T} \overline{x(u)} x(t+u)du.$$

This limit may exist pointwise a.e., weak* $\sigma(L^\infty, L^1)$, etc; for our purposes, if there is any doubt, the convergence is pointwise everywhere.

Proposition 2.2.   The autocorrelation P of $x \in L^\infty(\mathbb{R})$ is positive definite.

Proof.   i. Writing $\chi_T = H_{-T} - H_T$ we see that $(\chi_T x)*(\chi_T x)\tilde{}$ is positive definite since

$$\Sigma c_j\bar{c}_k(\chi_T x)*(\chi_T x)\tilde{}\,(t_j - t_k) - \int \Sigma c_j\bar{c}_k(\chi_T x)(u)\overline{(\chi_T x)(u + t_k - t_j)}du$$

$$= \int \Sigma c_j \overline{c_k} (\chi_T x)(w+t_j) \overline{(\chi_T x)(w+t_k)} du = \int |\Sigma c_j (\chi_T x)(u+t_j)|^2 du > 0.$$

ii. We can then verify that

$$\lim_{T \to \infty} \frac{1}{2T} \left(\int_{-T}^{T} \overline{x(u)x(t+u)} du - (\chi_T x)*(\chi_T x)\tilde{}(t) \right) = 0$$

for $t \in \mathbb{R}$.

iii. The result follows since limits of positive defin-ite functions are positive definite.

q.e.d.

The power spectrum S of the "sample function" $x \in L^\infty(\mathbb{R})$ with autocorrelation P $\gg$ 0 is the positive measure for which $\hat{S} = P$ a.e. For a given $x \in L^\infty(\mathbb{R})$ with power spectrum S and autocorrelation P we have the rela-tion

$$\int P(t)f*\tilde{f}(t)dt = \int |\hat{f}(\omega)|^2 dS(\omega) = \lim_{T \to \infty} \frac{1}{2T} \int_{-T}^{T} |f*x(t)|^2 dt \qquad (2.7)$$

for all $f \in L^1(\mathbb{R})$.

__Example 2.2.__ a. Let $x(t) = e^{it^2}$. Then $P(t) = 0$ if $t \neq 0$ and $P(0) = 1$, and the power spectrum $S = 0$.

b. If $x(\pm\infty) = 0$ or, more generally, if $\lim(1/(2T)) \int_{-T}^{T} |x(t)|^2 dt = 0$, then $P(t) = 0$ for all t and $S = 0$.

__Example 2.3.__ The value of an autocorrelation P is that it can be measured in many cases where the underlying signal x can not be quantified; the discrete part of the power spectrum S characterizes periodicities in x, e.g., [30, Chapter X] or the Couette flow experiments in fluid mechanics dealing with the onset of turbulence.

a. Let $x(t) = Ae^{2\pi it\omega}$. Then $\hat{S}(t) = P(t)$ is

$$P(t) = \lim_{T \to \infty} \frac{1}{2T} \int_{-T}^{T} \overline{A}e^{-2\pi iu\omega} Ae^{2\pi i(t+u)\omega} du = |A|^2 e^{2\pi it\omega},$$

and so $S = |A|^2 \delta_{-\omega}$.

b. The spectral analysis of a beam of light x(t) can
be made by a Michelson interferometer in the following way.
The intensity or power of the beam is the energy flow per
unit time (assuming area normalization) and is measured by a
power-sensitive photometer. The interferometer allows the
beam to take paths of different length so that the intensity
of x(u) + x(u+t) can be measured for various lags t.
Thus, the left-hand side of the equation,

$$\int |x(u) + x(u+t)|^2 du - 2\int |x(u)|^2 du = 2\int x(u)x(t+u)du,$$

can be measured, and hence the "autocorrelation" P(t) =
$\int$ x(u)x(t+u)du is computable.

3. STOCHASTIC PROCESSES

A $\underline{probability}$ $\underline{measure}$ on $\mathbb{R}$ is a positive measure p
ϵ M($\mathbb{R}$) for which there is an increasing surjection f:
$\mathbb{R} \cup \{\pm\infty\} \to [0,1]$ with the properties

   i. $\lim_{u \to t+} f(u) = f(t)$, for all $t \epsilon \mathbb{R}$,

   ii. $f' = p$, distributionally.

f is a $\underline{distribution}$ $\underline{function}$.

A $\underline{probability}$ $\underline{measure}$ on $P = \{\alpha_n: n \epsilon \mathbb{Z}\}$ is a function
p defined on the set $\mathbb{P}$ of all subsets of P for which
there is a sequence $\{p_n: p_n \geq 0$ and $\Sigma p_n = 1\}$ with the
property,

$$\forall A \epsilon \mathbb{P}, \ p(A) = \sum_{a_n \epsilon A} p_n.$$

$\underline{Remark}$ $\underline{3.1.}$ Probability measures p can be defined on any
space P. The pair (P,p) is a $\underline{probability}$ $\underline{space}$. In such
cases, if x: P $\to$ $\mathbb{R}$ is a "random variable" then there is a
probability measure p_x on $\mathbb{R}$ ($p_x(A) = p(x^{-1}(A))$) which is
the $\underline{distribution}$ $\underline{of}$ x $\underline{on}$ P. The distribution function f_x

of p_x is characterized by the property,

$$f_x(t) = p\{\alpha \ \varepsilon \ P : x(\alpha) \leqslant t\}.$$

To fix ideas we take $P = \mathbb{R}$ although this restriction is not necessary, as observed in Remark 3.1. A random variable x on $P = \mathbb{R}$, for a given probability measure p, is a real-valued pointwise p a.e. limit of a sequence of continuous functions. The mean of x, written $E\{x\}$, is

$$E\{x\} = \int_P x(\alpha)dp(\alpha) = \int_{-\infty}^{\infty} tdf_x(t) = \int_P x(\alpha)df(\alpha),$$

where the last two integrals are Stieltjes and where the last integral makes sense for $P = \mathbb{R}$ and f the distribution function of p.

Let (P,p) be a probability space. A function $x: \mathbb{R} \times P \to \mathbb{C}$ is a stochastic process if $x(t,.) : P \to \mathbb{C}$ is a random variable for each $t \ \varepsilon \ \mathbb{R}$. The expected value of the process x is $E\{x(t)\} = m(t)$. Suppose $x(t,.) \ \varepsilon \ L_p^2(P)$ for each $t \ \varepsilon \ \mathbb{R}$. Then x is a stationary stochastic process (SSP) if $m(t)$ is a constant m and if

$$\forall t,u,h \ \varepsilon \ \mathbb{R}, \ E\{x(t+h)\overline{x(u+h)}\} = E\{x(t)\overline{x(u)}\}.$$

We shall deal exclusively with SSPs x equipped with the continuity property , $\lim_{t\to 0} E\{|x(t)-x(0)|^2\} = 0.$

The autocorrelation of the SSP x is the continuous positive definite function

$$R(t) = E\{x(u+t)\overline{x(u)}\};$$

and the power spectrum of x is the positive measure S for which $\hat{S} = R$. The autocovariance of x is $C(t) = E\{(x(u+t)-m)\overline{(x(u)-m)}\} = R(t) - |m|^2$. The variance of each random variable $x(t,.)$ is $\sigma^2 = C(0) = R(0) - |m|^2$, and σ is the standard deviation. Note that $R(t) = E\{x(t)\overline{x(0)}\}.$

Example underline{3.1} Let x be a SSP. We compute

$$E\{x(t_1)\overline{x(t_2)}\} = E\{x(t_2+u)\overline{x(t_2)}\} = R(u),$$

for $u = t_1 - t_2$; and, hence, we have

$$E\{x(t_1)\overline{x(t_2)}\} = \int_{-\infty}^{\infty} e^{-2\pi i(t_2-t_2)\gamma} \, dS(\gamma).$$

Thus the mapping, $e^{2\pi i t \gamma} \rightarrow x(t,\alpha)$, establishes an isomorphism between

$$L_S^2(\mathbb{R}) = \{F : \|F\|_{2,S} = (\int |F|^2 dS)^{1/2} < \infty\}$$

and the span of the set $\{x(t,\cdot) : t \in \mathbb{R}\}$ completed by the norm

$$(\int_P |y(\alpha)|^2 dp(\alpha))^{1/2}.$$

Suppose the statistics of x are specified by the autocorrelation R. Then this isomorphism allows us to <u>predict</u> the process x at time $T > 0$ in terms of past knowledge of x (that is, for all $t < 0$) if $\exp(2\pi i T \gamma)$ belongs to the closure in $L_S^2(\mathbb{R})$ of the span of the set $\{\exp(2\pi i t \gamma): t < 0\}$.

<u>Example 3.2</u> If x is a SSP with autocorrelation R then $\overline{E\{|x(t+u) \pm x(u)|^2\}} = 2\text{Re}(R(0) \pm R(t))$, cf., Example 2.3.

In order to reconcile the two definitions of autocorrelation we make the following definition. A SSP x is a <u>correlation ergodic process</u> if

$$\forall t \in \mathbb{R}, \lim_{T \to \infty} \frac{1}{2T} \int_{-T}^{T} x(t+u,\alpha)\overline{x(u,\alpha)}du = R(t)$$

in measure. (Recall that $\lim_T f_T(\alpha) = 0$ in <u>measure</u> or in <u>probability</u> if for $\forall \varepsilon > 0$, $\lim_T p\{\alpha: |f_T(\alpha)| > \overline{\varepsilon}\} = 0$. Also, if $\lim_T f_T(\alpha) = 0$ in measure then there is a subsequence $\{T_n\}$ for which $\lim_n f_{T_n}(\alpha) = 0$ a.e.)

<u>Proposition 3.1.</u>   Let x be a SSP with autocorrelation R. x is a correlation ergodic process if

$$\forall t \in \mathbb{R}, \lim_{T \to \infty} \frac{1}{2T} \int_{-T}^{T} C(t,v)(1 - \frac{|v|}{2T})dv = 0, \tag{3.1}$$

where $C(t,v) = E\{x(t+u+v)\overline{x(u+v)}x(t+u)\overline{x(u)}\} - |R(t)|^2$, cf., [21, pp. 354-360].

Proof.   Let $P_T(t,\alpha) = (1/(2T)) \int_{-T}^{T} x(t+u,\alpha)\overline{x(u,\alpha)}du.$

Clearly, $E\{P_T(t,.)\} = R(t)$, and a routine calculation shows that the variance $E\{|P_T(t,.) - R(t)|^2\}$ of P_T is the term (without the "lim") on the left-hand side of (3.1).

The result follows from (3.1) and Tchebyshev's inequality,

$$p\{\alpha: |P_T(t,\alpha) - R(t)| > \varepsilon\} < \frac{1}{\varepsilon^2} \int_P |P_T(t,\alpha)-R(t)|^2 dp(\alpha).$$

q.e.d.

Therefore, to establish correlation ergodicity of x we need knowledge of its fourth order moments.

Remark 3.2   Let us mention two other processes besides correlation ergodic processes.

a. **Ergodic processes** x, where **all** statistical properties are determined by one sample path $x(.,\alpha)$, are uninteresting in spectrum estimation since the power spectra of such processes have no discrete part, e.g., [8, pp. 76-78].

b. **Stationary white noise** w on $\mathbb{R}$ is characterized by the property that its autocorrelation $R_w = I\delta$, where I > 0 is an intensity constant. R_w is **positive definite** in the sense that

$$\forall \phi \; \varepsilon \; C_c^\infty(\mathbb{R}), \; \langle I\delta, \; \phi*\tilde{\phi}\rangle > 0;$$

in fact, $\langle I\delta, \; \phi*\tilde{\phi}\rangle = I \; \|\phi\|_2^2$, cf., (2.6).   The power spectrum S_w of w is the constant function $S_w(\omega) = I$.

Remark 3.3.   White noise is not a stochastic process but can be formulated precisely as a generalized stochastic process w: $C_c^\infty(\mathbb{R}) \to L^2(P)$, where $(P,\mathbf{P},p)$ is a probability space and where we have replaced $\mathbb{R}$ by $C_c^\infty(\mathbb{R})$.   In fact, w is the (Schwartz) distributional derivative of the Wiener process (Brownian motion), e.g., [10, Sections III.1 and III.2], cf., [2, pp. 529-533; 5]. **Digital stationary white noise** is much easier to formulate. It is a stochastic process $w(k,\alpha)$ consisting of a sequence of uncorrelated random variables with the stationarity constraints, $E\{w(k)\} = m$

and $E\{|w(k)-m|^2\} = \sigma^2$ for all k, e.g., [25, Section 3.5.1].

4. PERIODOGRAMS

Given $b \varepsilon L^1(\mathbb{R})$ and suppose x is a SSP such that each sample function $x(.,\alpha)$ is an element of $L^\infty(\mathbb{R})$. Then

$$S_b(\omega,\alpha) = |\int b(t)x(t,\alpha)e^{-2\pi i t\omega}dt|^2$$

is the <u>periodogram</u> associated with the process x and <u>data window</u> b.

Proposition 4.1.   Let x be a SSP and suppose $b \varepsilon L^1(\mathbb{R})$. If S is the power spectrum of x then

$$E\{S_b(\omega)\} = \int_{-\infty}^{\infty}|B(\omega+\gamma)|^2 dS(\gamma), \quad b \leftrightarrow B. \qquad (4.1)$$

If x is real and $b = \tilde{b}$ then

$$E\{S_b(\omega)\} = S*B^2(\omega).$$

Proof.

$$E\{S_b(\omega)\} = \int\int b(t)\overline{b(u)}e^{-2\pi i(t-u)\omega}(\int_P x(t,\alpha)\overline{x(u,\alpha)}dp(\alpha))dtdu.$$

Writing $\underset{\approx}{x}(t,\alpha) = x(u+(t-u),\alpha)$ we see that $\int_P$ is $R(t-u) = \hat{S}(t-u)$ and so $E\{S_b(\omega)\}$ becomes

$$\int[\int\int b(t)\overline{b(u)}e^{-2\pi i(t-u)(\omega+\gamma)}dtdu]dS(\gamma),$$

and this is (4.1).   q.e.d.

Proposition 4.2.   Let x be a real SSP and suppose $b_T = \sqrt{\pi}\,\chi_T$. If S is the power spectrum of x then

$$\lim_{T\to\infty} E\{S_{b_T}(\omega)\} = S \qquad (4.2)$$

in the $L^1(\mathbb{R})$ weak * sense, i.e.,

$$\forall f \in L^1(\mathbb{R}),\ f \leftrightarrow F,\ \lim_{T \to \infty} \langle S-E\{S_{b_T}(\omega)\},F\rangle = 0.$$

Because of (4.2) we can assert that S_{b_T} is an <u>asymp-</u><u>totically</u> <u>unbiased</u> <u>estimator</u>. Any data window $b_T \leftrightarrow B_T$, supported by $[-T,T]$, can be used in Proposition 4.2 as long as B_T is an approximate identity,

i.e., $\int B_T^2 = 1$, $\sup_T \int |B_T^2| < \infty$, and $\lim_T \int_{|\omega|>\varepsilon} |B_T(\omega)|^2 d\omega = 0$ for each $\varepsilon > 0$.

5. THE SPECTRUM ESTIMATION PROBLEM

The spectrum estimation problem is to clarify and quantify the statement: find periodicities in a signal x recorded over the fixed time interval $[-T, T]$. In more picturesque language, we want to filter the noise from the incoming signal x in order to determine the intelligent message (periodicities) therein.

We assume the following mathematical model for the frequency deconvolution (FDA) algorithm presented in Section 6.

<u>Assumptions</u>.   1. The signal x is actually defined on the product space $[-T,T] \times P$, where P is a probability space and x is the restriction to $[-T,T] \times P$ of some SSP y.

2. The expectation E_b of the periodogram $S_b(\omega,\alpha)$ is known, where $\text{supp } b \subseteq [-T,T]$ and $\|b\|_2 = 1$.

3. The power spectrum S of x is uniquely determined. This assumption is a theorem in case we make the experimentally reasonable hypothesis that the power spectrum S_y of every stationary extension y of x is supported by $[\Sigma_\Omega,\infty)$ (<u>JMAA</u> 91(1983) 444-509, Proposition III.1.3, and Section 1 of "Fourier uniqueness criteria and spectrum estimation theorems" in this volume), cf. [26].

<u>Remark 5.1</u>.   a. Assumption 3 is not universally accepted. In fact, the maximum entropy method (MEM) of spectrum estimation, which we discuss in Section 7, is based on a point of

view opposite that of a uniquely determined power spectrum.
MEM does not assert the existence of a unique power spectrum
and then estimate it; instead, given x on [-T,T] × P or
the autocorrelation R on [-T,T], MEM models autocorrelation data outside [-T,T] by maximizing a certain entropy
integral.

b. Besides FDA (Section 6) and MEM (Section 7) there
are many other estimators which, roughly speaking, are windowing methods (e.g., FDA) or high resolution methods (e.g.,
MEM). The choice of method frequently depends on the problem
at hand and we refer to [3; 14b; 15] for lucid discussions of
this point. Technical presentations of classical windowing
methods are found in [1; 20, Chapter 11; 21, Chapter 12; 25];
since such methods deal with periodograms they are useful
because of the efficiency of the fast Fourier transform
(FFT). Technical presentations of high resolution methods
are found in [2, pp. 511-512; 4; 9; 14; 15; 19; 22; 24].

c. Assumption 1 preempts all genuine statistical problems, e.g., [2, pp. 509-510; 21, Chapter 12]; and FDA does
not completely come to grips with reduction of variance, cf.,
JMAA 91 (1983) 444-509, Theorem III.4.1.

Example 5.1.   If x is a SSP and $E\{x\} = m \neq 0$ then we
can verify that the power spectrum S_x of x is

$$S_x = A\delta + \mu, \tag{5.1}$$

where $A > 0$ and μ is a measure for which $\mu(\{0\}) = 0$. To
see this we first compute the autocovariance

$$R_y(t) = E\{\overline{(x(u)-m)}(x(u+t)-m)\} = R_x(t)-|m|^2 \tag{5.2}$$

and set $y(t,\alpha) = x(t,\alpha)-m$. y is a SSP, $E\{y\} = 0$, and
$S_y = S_x-|m|^2\delta > 0$ (by (5.2)). Writing S_y as $S_y(\{0\})\delta+\mu$,
where $S_y(\{0\}) > 0$ and $\mu(\{0\}) = 0$ since

$S_y > 0$, we have

$$S_x = S_y + |m|^2\delta = (|m|^2+S_y(\{0\}))\delta+\mu$$

and so

$$A = |m|^2 + S_y(\{0\}) > 0.$$

A similar calculation is valid for a sample function
x(t) with non-zero mean,

$$m = \lim(1/(2T))\int_{-T}^{T} x(t)dt.$$

In practice it is frequently important to remove the mean from the data since the corresponding peak in the power spectrum can hinder our ability to see the rest of the spectrum.

6. FDA

Given the <u>Assumptions</u> of <u>Section 5</u> for real valued x defined on $[-T,T] \times P$ and for a data window $b = \tilde{b} \leftrightarrow B$ supported by $[-T,T]$ and with norm $\|\|b\|\|_2 = 1$.   Then

$$E\{S_b(\omega)\} = S*B^2(\omega), \qquad (6.1)$$

where $E_b = E\{S_b\}$ and B^2 are known.

FDA$_2$ is formulated as follows for $B^2(0) \neq 0$. Set $C = (1/B^2(0))B^2 H$.   For any $c > 0$ we define

$$C_c = \sum_{n=0}^{\infty} C(nc)\chi_{[nc,(n+1)c]}.$$

Thus, $C_c * C_c^{-1} = \delta$ for

$$C_c^{-1} = \sum_{n=0}^{\infty} a_n \delta'_{nc}$$

where $a_0 = 1$ and

$$\forall n > 1, \; a_n = 1 - \sum_{m=0}^{n-1} C((n-m)c)a_m.$$

Applying C_c^{-1} to both sides of (6.1) and renormalizing by $B^2(0)$ we are able to define the <u>B-estimator</u> of S, with sampling length c, as

$$\text{(FDA)} \qquad S_{bc} = \frac{1}{B^2(0)} \sum_{n=0}^{\infty} a_n \delta'_{nc} * E_b.$$

A similar algorithm can be formulated for non-uniformly semi-group distributed spectral data (<u>SIAM</u> <u>J</u>. <u>Math</u>. <u>Analysis</u> 13 (1982) 180-207).

FDA is only effective if b is chosen to minimize the
difference between the unknown but uniquely defined S and
the computable S_{bc}. The following motivation and example
show how this can be done.

Motivation.   a. Suppose that   supp S $\subseteq$ [-Ω,Ω]   and that
$B = \sqrt{1/\Gamma}\chi_{[0,\Gamma]}$.   (By analyticity, B cannot be exactly of
this form since   supp b $\subset$ [-T,T].)   In this case E_b is an
increasing function on the frequency interval   [-Ω,Ω]   as
long as $\Gamma > 2\Omega$; and, further, the small frequency intervals
of greatest increase for E_b contain the frequencies at which
the peaks of S occur. We quantify this picture in Example
6.1.

       b. If B^2 is an approximate identity, as in Proposi-
tion 4.2, then E_b also has a pictorial representation which
allows one to see the peaks. In fact, the picture will be
that of a bumpy curve of hills and dales, where the hills
indicate the peaks of S. An advantage of the increasing
curve E_b in part a is that deconvolution can be implemented
to produce an unbiased estimator. In practice, when dealing
with approximate identities, bias is reduced by tapering to
correct for leakage into the side-lobes of B^2, cf., Example
6.3.

Example 6.1.   Given S,Γ, and $B = \sqrt{1/\Gamma}\ \chi_{[0,\Gamma]}$ as in the
Motivation.

       a. The inverse of $\chi_{[0,\Gamma]}$ under convolution is

$$\chi_{[0,\Gamma]}^{-1} = \sum_{n=0}^{\infty} \delta_{n\Gamma}$$

and so the B-estimator of S is

$$S_{b\Gamma}(\omega) = \Gamma \sum_{n=0}^{\infty} E_b(\omega - n\Gamma),$$

which yields $S_{b\Gamma} = \Gamma E_b$ on $(-\Omega,\Omega)$ for $\Gamma > 2\Omega$.

       b. Suppose that S consists of moderately white noise
S_N in [-Ω,Ω] plus the sum

$$S_P = \sum_{k=-K}^{K} m_k \delta_{\omega_k}, -\Omega < \omega_{-K} < \ldots < \omega_{-1} < \omega_0 = 0 < \omega_1 < \ldots < \omega_K < \Omega,$$

of peaks having masses $m_{|k|} > 0$ at ω_k. Then $S_N > 0$ is relatively constant on $[\frac{1}{2}\Omega, \Omega]$ and

$$E_b(\omega) = S_N * B^2(\omega) + \sum_{k=-K}^{K} m_k B^2(\omega - \omega_k).$$

Thus, E_b jumps about m_{-K}/Γ at ω_{-K} and remains relatively constant from there until ω_{-K+1} at which point it jumps another m_{-K+1}/Γ. Consequently, the jump of E_b at ω_k is m_k/Γ; and so the peak at ω_k can be detected if m_k/Γ is large vis a vis the noise. If B is realizable (and, thus, not

$$\sqrt{1/\Gamma} \chi_{[0,\Gamma]})$$

then this example can even be used to detect peaks which are within $1/T$ of each other if some information about ω_k or m_k or S_N is also known, e.g., Example 6.2b.

Example 6.2.   a. The fundamental theoretical source of error between S_{bc} and S arises from considering HB^2 instead of B^2. In this regard note that $|B'(\omega)|$ is always bounded by

$$2\pi\sqrt{2/3} \ T^{3/2} \|b\|_2 \quad \text{for} \quad \text{suppb} \subseteq [-T,T].$$

This estimate provides restrictions on how steeply B^2 can increase from (about) 0 to (about) $1/\Gamma$ on $(-\infty,0]$, thus motivating the need of prior knowledge as suggested in Example 6.1b.

    b. On the other hand, two close peaks can theoretically be detected as follows. Suppose we are given the setup of Example 6.1b where $K = 1$, S_N is very flat, $\omega_{-1} = 0$, and $|\omega_0 - \omega_1|$ is much smaller than $1/T$ for the fixed data length $2T$. Choose B of the form

$$B = c\tau_\Gamma \frac{1}{\sqrt{\Gamma}} \chi_\Gamma * \hat{\chi}_\Gamma.$$

Thus B will be about $1/\sqrt{\Gamma}$ locally for positive frequencies and will drop off to 0 at a negative frequency near $-1/T$.

Because we are dealing with the sum, $s = m_0 B^2(\omega)$
$+ m_1 B^2(\omega-\omega_1)$, the idea is to look at the intervals
$I_0 = [-1/T,\omega_1-1/T]$, $J_0 = [\omega_1-1/T,\omega_0]$, and $_2I_1 = [\omega_0,\omega_1]$. On
I_0 s will be a product of a translate of B^2 times $\tilde{m}_0/\Gamma$,
on I_1 it will be a product of a translate of B^2 times m_1/Γ
plus the constant m_0/Γ, and on J_0 it will be a sum of two
different translates of B^2. Since B^2 is explicit on
$[-1/T,0]$, we can distinguish between the graphs of such products and sums, the former flanking the latter on three con-
secutive small intervals. Consequently, if such behaviour is
detected in a small region where we know there is a peak then
we can conclude that there are two close peaks. When noise
is added it has the effect of placing an error margin about
such products and sums (on I_1 and J_0, respectively), and
so two close peaks can be detected by this method if the factors m_0/Γ and m_1/Γ are large enough.

The same method could be used for approximate identities
B^2 but the picture becomes more difficult to read because of
loss of monotonicity.

Example 6.3.   Consistent estimators

a. For a given T and a given restriction to
$[-T,T] \times P$ of a SSP x, let $\tilde{S}_T$ be a continuous spectrum
estimator of the power spectrum S. The **bias** of the estimator is

$$b_T = S - E\{\tilde{S}_T\}$$

and the **variance** of the estimator is

$$var_T = E\{|\tilde{S}_T - E\{\tilde{S}_T\}|^2\} = E\{|\tilde{S}_T|^2\} - |E\{\tilde{S}_T\}|^2.$$

Both of these notions depend on frequency points ω. The
estimator S_T is <u>consistent</u> if

$$\lim_{T\to\infty} b_T = 0 \quad \text{and} \quad \lim_{T\to\infty} var_T = 0.$$

(This definition depends on the type of convergence, e.g.,
pointwise, and on the structure of S in case S is not a
smooth function.)

b. Let T be written as a product $T = T_b T_v$. Choose a
real <u>data window</u> $b = b_T \leftrightarrow B_T$ such that

supp $b \subseteq [0, T_b]$, $\|b_T\|_2 = 1$, and $\{B_T^2\}$ is an approximate identity. Assume T_v is a positive integer N and let b_j be the translate $\tau_{jT_b} b$ of b for $j = 0, 1, \ldots N-1$. Thus,

supp $b_0 \subseteq [0, T_b]$, supp $b_1 \subseteq [T_b, 2T_b], \ldots,$ supp $b_{N-1} \subseteq [(N-1)T_b, NT_b]$.

Also, for each j, we define the periodogram

$$S_{b_j}(\omega, \alpha) = \left| \int_{jT_b}^{(j+1)T_b} b_j(t) x(t, \alpha) e^{-2\pi i t \omega} dt \right|^2.$$

In order to construct a consistent estimator we must make the <u>assumption</u> that $E\{S_{b_j}\} = E\{S_b\}$ for each j.

We define the <u>Bartlett</u> estimator $\tilde{S} = \tilde{S}_T$ as

$$\tilde{S}(\omega, \alpha) = \frac{1}{N} \sum_{0}^{N-1} S_b(\omega, \alpha).$$

The expectation of $\tilde{S}_T$ is

$$E\{\tilde{S}(\omega, .)\} = \frac{1}{N} \sum_{0}^{N-1} E\{S_{b_j}(\omega, .)\} = E\{S_b(\omega)\},$$

where the last term follows from our assumption that the expected value of S_{b_j} is the same as that of S_b. Thus,

$$E\{\tilde{S}_T(\omega, .)\} = S * B_T^2(\omega),$$

and so $\tilde{S}_T$ is an asymptotically unbiased estimator in the case $\{B_T^2\}$ <u>is an approximate identity</u>, assuming we let $T_b \to \infty$ when $T \to \infty$.

In order to compute the variance of $\tilde{S}_T$ we first compute

$$E\{|\tilde{S}(\omega)|^2\} = \frac{1}{N^2} \sum_{j=0}^{N-1} \sum_{i=0}^{N-1} E\{S_{b_j}(\omega, .) \overline{S_{b_i}(\omega, .)}\} =$$

$$\frac{1}{N^2} \sum_{j=0}^{N-1} E\{|S_{b_j}(\omega,.)|^2\} + \frac{1}{N^2} \sum_{i \neq j} E\{S_{b_j}(\omega,.) \overline{S_{b_i}(\omega,.)}\},$$

which, by our **assumption**, can be written as

$$\frac{1}{N}(\frac{1}{N} \sum_{j=0}^{N-1} E\{|S_{b_j}(\omega)|^2\}) + \frac{1}{N}(\frac{1}{N} \sum_{i \neq j} E\{S_{b_j}(\omega)\}E\{\overline{S_{b_i}(\omega)}\}) =$$

$$\frac{1}{N} E\{|S_b(\omega)|^2\} + \frac{1}{N}(\frac{1}{N} \sum_{i \neq j} E\{S_{b_j}(\omega)\}E\{\overline{S_{b_i}(\omega)}\}) =$$

$$\frac{1}{N} E\{|S_b(\omega)|^2\} + \frac{1}{N}(\frac{1}{N} \sum_{i \neq j} |E\{S_b(\omega)\}|^2).$$

Clearly, $\sum_{i \neq j} 1 = N(N-1)$, which is the number of terms in an $N \times N$ matrix less the N diagonal elements. Therefore, we have

$$E\{|\tilde{S}(\omega,.)|^2\} = \frac{1}{N} E\{|S_b(\omega)|^2\} + \frac{N-1}{N}|E\{S_b(\omega)\}|^2.$$

Consequently, we see that

$$\text{var } \tilde{S}_T(\omega) = E\{|\tilde{S}(\omega)|^2\} - |E\{\tilde{S}(\omega)\}|^2 =$$

$$\frac{1}{N} E\{|S_b(\omega)|^2\} - \frac{1}{N}|E\{S_b(\omega)\}|^2 = \frac{1}{N} \text{var } S_b(\omega);$$

and so $\lim_{T \to \infty} \text{var } \tilde{S}_T(\omega) = 0$, assuming $N = T_v \to \infty$ as $T \to \infty$.

Thus, the **Bartlett estimator** $\tilde{S}_T$ **is a consistent estimator**.

   c. Along with Bartlett's consistent estimator we also mention the related work of Blackman and Tukey. Generally, one considers estimators $S*B*W$, for $b \leftrightarrow B$ and $w \leftrightarrow W$, where the convolution by W replaces the arithmetic mean in the definition of the Bartlett estimator. The temporal window w is referred to as a **lag window**.

   A reason for developing estimators beyond the periodogram is that S_b is **not** a good estimator of S, even for large T, since the variance can't be made small. In order to reduce the variance of the estimator we are lead to deal with estimators of _smoothed_ versions of S. In particular _we reduce resolution_. This playoff between variance and resolution can be quantified into an "uncertainty principle", e.g. Ark. Mat. 1(1951) 503-531, especially page 524 (by U. Grenander).

Example 6.4.   Related to the problem of detecting two close peaks we mention the following fact which shows how number theory can arise. Suppose x is a stationary process with real autocorrelation R. It is well-known that $R(0) > |R(t)|$ and that if $|R(t)| = R(0)$ for some t then $R(u)=\exp(iu\alpha/t)p(u)$ for some $\alpha \in \mathbb{R}$, where p is t-periodic, e.g. [21,p.315]. Thus, if $R(t_1) = R(t_2) = R(0)$ and $t_1/t_2 \notin \mathbb{Q}$ then $S = R(0)\delta$.

7. MEM

In order to provide background for the method of this section we distinguish two aspects of the spectrum estimation problem [22]:

Deterministic problem.   Estimate the power spectrum of a SSP x in terms of given autocorrelation data $R(T)$, t $\in$ [-T,T];

Stochastic problem.   Estimate the power spectrum of a SSP x in terms of a given sample function $x(t,\alpha)$, α fixed and t $\in$ [-T,T].

Neither of these problems is really meaningful mathematically since many power spectra can be constructed which satisfy these conditions. The physical point of view which allows a solution to these problems, especially in the context of entropy, is persuasively presented in [15]. The solution is a power spectrum S which maximizes the entropy rate, $\int \log S(\omega)d\omega$, of the process. Loosely speaking, this maximization of entropy is a mathematical guarantee that the least number of assumptions have been made regarding the information content of the unmeasured data at $|t| > T$, e.g., [4, pp. 94-96 (J. Edward and M. Fitelson)].

We shall solve the <u>Deterministic problem</u>; the solution
is actually a theorem (Theorem 7.1). The practical reliabil-
ity of this MEM theorem depends on the real meaning of
entropy for a given problem; and so there are the same questions about applicability and usefulness of MEM as there are
about algorithms such as FDA. In fact, MEM has been known to
exhibit peaks which don't exist and there are still questions
about its effectiveness for non-uniformly distributed data
and in higher dimensions, cf., [14c].

The proof we give of the MEM theorem is due to Dym and
Gohberg; there is a long history of the method, and its
derivations and applications, in the engineering and statistical literature beginning with Burg's thesis. We give the
discrete version [6] since the analog version [7] is more
complicated to formulate.

A matrix $R = (r_{ij})$, $r_{ij} \varepsilon \mathbb{C}$ and $i,j = 0,\ldots,n$,

is <u>hermitian</u> if $r_{ij} = \overline{r_{ji}}$. An hermitian matrix R is <u>posi-
tive</u> if

$$\langle Rc,c \rangle = \sum_{i,j} r_{ij} c_j \overline{c_i} \geqslant 0$$

for each $c \varepsilon \mathbb{C}^{n+1}$. A positive hermitian matrix R is
<u>positive definite</u>, written $R \gg 0$, if $\langle Rc,c \rangle = 0$ implies
$c = 0$. Thus, positive matrices correspond to positive defin-
ite functions, whereas positive definite matrices have the
additional uniqueness property that $\langle Rc,c \rangle > 0$ for $c \neq 0$.

<u>Remark 7.1.</u> Given a matrix R with eigenvalues $\{\lambda_0,\ldots,\lambda_n\}$.
Then $\sum r_{ii} = \sum \lambda_i$ and $\det R = \Pi \lambda_i$. If R is hermitian then
it is a standard result in linear algebra that $R \gg 0$ if and
only if each $\lambda_j > 0$.

<u>Example 7.1.</u>   Given $\{r_j: j = 0, \pm 1,\ldots,\pm n$ and $\overline{r_j} = r_{-j}\} \subseteq$
$\mathbb{C}$. Let $R = (r_{ij}) = (r_{i-j})$, the "Toeplitz" matrix constant
on "diagonals of negative slope". R is hermitian since
$\overline{r_j} = r_{-j}$. By Remark 7.1 we see that $\det R > 0$ if $R \gg 0$.
In this case we also have $\sum \lambda_j = (n+1)r_0$ and hence $r_0 > 0$;
also $R^{-1} = (c_{ij})$ exists since $\det R \neq 0$.

The key to the MEM theorem is the following classical result due to Szegö, e.g., Toeplitz forms and their applications by U. Grenander and G. Szegö.

Lemma 7.1.   Given $\{r_j: j = 0,\pm 1,\ldots,\pm n$ and $\bar{r}_j = r_{-j}\}$ and assume the matrix $R = (r_{i-j}) \gg 0$. The unique column matrix solution $A = (a_0,\ldots,a_n)$ of the matrix equation $RA = (\delta_{j1})$ has the property that

$$\forall \omega, \quad A(e^{2\pi i \omega}) = \sum_{0}^{n} a_j e^{2\pi i j \omega} \neq 0.$$

(δ_{j1} is 1 or 0, for $j = 1,\ldots,n+1$, depending on whether $j \neq 1$ or not.)

We also use the following classical result due to Wiener, e.g., The Fourier integral and certain of its applications by N. Wiener.

Lemma 7.2.   Let $A(e^{2\pi i \omega})$ be a non-vanishing absolutely convergent Fourier series. Then $A^{-1} = 1/A$ is an absolutely convergent Fourier series.

Theorem 7.1 (MEM).   Given $\{r_j: j = 0,\pm 1,\ldots,\pm n$ and $\bar{r}_j = r_{-j}\}$ and assume $R = (r_{i-j}) \gg 0$. There is a unique absolutely convergent Fourier series $S \sim \Sigma\, s_j e^{2\pi i j \gamma}$, $\hat{S}(j) = s_j$, with the following properties:

a. $\forall |j| < n, \quad s_j = r_j$;

b. $S(e^{2\pi i \gamma}) > 0$ for all γ (and hence S^{-1} is an absolutely convergent Fourier series by Lemma 7.2);

c. $S = |S_+|^2$ where $S_+ \sim \sum_{0}^{\infty} s_j^+ e^{2\pi i j \gamma}$ is an absolutely convergent Fourier series;

d. $\forall |j| > n, \quad (S^{-1})^{\wedge}(j) = 0$ and

$$S(e^{2\pi i \gamma}) = 1/[\sum_{k=-n}^{n} (\frac{1}{c_{00}} \sum_{j=0}^{n} \overline{c}_{0j} c_{0,j-k}) e^{2\pi i k\gamma}],$$

where $R^{-1} = (c_{ij})$;

    e. For all absolutely convergent Fourier series $G > 0$, for which $\hat{G}(j) = r_j$ when $|j| < n$, we have
$$\int_0^1 \log G d\gamma < \int_0^1 \log S d\gamma;$$
and equality is obtained if and only if $G = S$.

Proof.   Using the hypothesis $R \gg 0$, choose the non-vanishing trigonometric polynomial A from Lemma 7.1. It is easy to see that $a_0 = c_{00} > 0$. By Lemma 7.2 we see that $1/A$ is an absolutely convergent Fourier series; and since the set of all such series is closed under pointwise multiplication we have that
$$S = a_0/|A|^2 > 0$$
is also an absolutely convergent Fourier series. We set $S_+ = \sqrt{a_0}/A$. At this point, except for some straightforward computations, parts $\underline{a} - \underline{d}$ of the result have been proved.

    Now take any positive absolutely convergent Fourier series G. We can write $G = G(0)|G_+|^2$, where
$$|G_+(0)| = 1, \quad G_+ \sim \sum_0^\infty g_j^+ e^{2\pi i j\gamma}, \quad \text{and} \quad \Sigma |g_j^+|^2 < \infty;$$
and it is a classical fact that
$$\int_0^1 \log|G_+| = \log|G_+(0)| = 1,$$
e.g., Section 2 of "Fourier uniqueness criteria and spectrum estimation theorems" in this volume. Consequently
$$\int_0^1 \log G d\gamma = \log G(0). \tag{7.1}$$
Also by our definition of S we have
$$\int_0^1 \log S d\gamma = \log a_0 - \int_0^1 \log|A|^2 d\gamma = \log a_0 - \log|A(0)| = \log a_0^{-1}. \tag{7.2}$$
We now write G as

$$G = S + (G-S) = A^{-1} a_0 \overline{A}^{-1} + (G-S)$$

and, hence, we obtain

$$a_0^{-1} A G \overline{A} a_0^{-1} = a_0^{-1} + a_0^{-1} A(G-S)\overline{A} a_0^{-1}. \tag{7.3}$$

The 0th Fourier coefficient of $a_0^{-1} A(G-S)\overline{A} a_0^{-1}$ is 0, and this is where we use the hypothesis that $\hat{G}(j) = \hat{S}(j)$ for $|j| < n$. Thus, the 0-th Fourier coefficient of the right hand side of (7.3) is a_0^{-1}. A direct computation of the 0-th Fourier coefficient of the left hand side of (7.3) yields

$$G(0)(1 + \sum_1^\infty |d_j|^2) > G(0).$$

Therefore, $a_0^{-1} > G(0)$. This inequality combined with (7.1) and (7.2) yields part e.

q.e.d.

Theorem 7.1d   reflects the fact that MEM provides an all-pole spectrum estimator. As such, the spectrum estimator S of Theorem 7.1 can be interpreted (e.g., Example 7.2) as the power spectrum of an autoregressive process (AR), cf., [4, pp. 99-118 (Makhoul)] for a discussion of the more general ARMA models.

Example 7.2.   An autoregressive process is the output x of a recursive linear system with a white noise forcing function. The continuous version of such a system is characterized by the differential equation,

$$\sum_{j=0}^n c_j x^{(j)}(t,\alpha) = w(t,\alpha),$$

and the system function H has the form

$$H(z) = I/(c_n z^n + \ldots + c_0).$$

Since $x = h*w$, $h \leftrightarrow H$, and since the power spectrum S_w of w is I, we have

$$S_x = |H(-\omega)|^2 S_w = I|H(-\omega)|^2;$$

and so the power spectrum S_x of x has the form given in Theorem 7.1d.

REFERENCES

1. R. Blackman and J.W. Tukey, "The measurement of power spectra," Dover, New York (1959).

2. D. Brillinger, "Time series," expanded edition, Holden-Day, San Francisco, (1981).

3. D. Brillinger and J.W. Tukey "Spectrum estimation and system identification relying on a Fourier transform".

4. D. Childers, ed., "Modern Spectrum Analysis," IEEE (1978) (many of the "classical" papers).

5. R. Dudley, Sample functions of the Gaussian process, Ann. of Prob.   1: 66-103(1973).

6. H. Dym and I. Gohberg, Extensions of matrix valued functions with rational polynomial inverses, Int. Eq. and Operator Theory 2: 503-528 (1979).

7. H. Dym and I. Gohberg, On an extension problem, generalized Fourier analysis, and an entropy formula, Int. Eq. and Operator Theory 3: 143-215 (1980).

8. H. Dym and H. McKean, "Gaussian processes, function theory, and the inverse spectral problem," Academic Press, New York, (1976).

9. P. Fougere, Spontaneous line splitting in multi-channel maximum entropy power spectra, in: "ASSP Workshop on Spectral Estimation," S.S. Haykin, ed., McMaster University, Hamilton (August, 1981).

10. I. Gelfand and N. Vilenkin, "Generalized functions, Volume 4," Academic Press, New York (1964).

11.   U. Grenander, ed., "The Cramer Volume," John Wiley & Sons, New York, (1959) (esp. the article by Tukey).

12.   B. Harris, ed., "Spectral analysis of time series," John Wiley & Sons, New York, (1966-1967) (esp. the introduction and the article by Tukey).

13.   F. Harris, On the use of windows for harmonic analysis, Proc IEEE 66: 51-83 (1978).

14.   IEEE Proceedings Volume 70 (Sept. 1982).

      a. E.A. Robinson, A historical perspective of spectrum estimation, Proc. IEEE 70: 885-907 (1982).
      b. E.T. Jaynes, On the rationale of maximum entropy methods, Proc. IEEE 70: 939-953 (1982).
      c. J. McClellan, Multidimensional spectral estimation, Proc. IEEE 70: 1029-1039 (1982).
      d. D.J. Thomson, Spectrum estimation and harmonic analysis, Proc. IEEE 70: 1055-1096 (1982).

15.   E.T. Jaynes, On the rationale of maximum entropy methods, in: "First ASSP Workshop on Spectral Estimation," S.S. Haykin, ed., McMaster University, Hamilton (August, 1981).

16.   J. Lamperti, "Probability" Benjamin Publishing, New York, (1966).

17.   J. Lamperti, "Stochastic processes" Springer-Verlag, New York, (1977).

18.   H. Landau, H. Pollak, D. Slepian, Prolate spherical wave functions..., Bell System Tech. J. 40: (1961) 43-64 (1961), 40: 65-84 (1961), 41: 1295-1336 (1962), 43: 3009-3058 (1964), 57 1371-1430 (1978).

19.   A. Oppenheim, ed., "Applications of Digital Signal Processing," Prentice-Hall, Inc., Englewood Cliffs, New Jersey, (1978).

20.   A. Oppenheim and R. Schafer, Digital Signal Processing, Prentice-Hall, Inc., Englewood Cliffs, New Jersey, (1975).

21. A. Papoulis, "Signal Analysis," McGraw-Hill, New York, (1977).

22. A. Papoulis, Maximum entropy and spectral estimation, a review, IEEE Trans ASSP 29 (6): 1176-1186 (1981).

23. A. Papoulis,"Probability, random variables, and stochastic processes," (there may be a new edition soon).

24. E. Parzen, Autoregressive spectral estimation,..., in: "First ASSP Workshop on Spectral Estimation," S.S. Haykin, ed., McMaster University, Hamilton (August, 1981).

25. M. Priestley, "Spectral analysis and time series," Volumes 1 and 2, Academic Press, New York, (1981).

26. D. Slepian, On bandwidth, Proc. IEFE 64: 292-300 (1976).

27. L. Schwartz, "Théorie des distributions," Hermann, Paris, (1966).

28. D.E. Vakman,"Sophisticated signals and the uncertainty principle in radar,"Springer Verlag, Berlin (1968).

29. N. Wiener, "Extrapolation, interpolation, and smoothing of stationary time series, with engineering application," MIT Press, Cambridge, Mass. (1949).

30. N. Wiener, "Cybernetics" 2nd ed., MIT Press (1961).

31. N. Wiener, "Generalized harmonic analysis and Tauberian Theorems," MIT Press, Cambridge, Mass. (1966). Also in Wiener's "Oeuvres, Vol. II," MIT Press.

32. N. Wiener Oeuvres, Volume III, MIT Press.

SOME PRACTICAL AND STATISTICAL ASPECTS OF

FILTERING AND SPECTRUM ESTIMATION

T.P. Speed

CSIRO Division of Mathematics and Statistics
Canberra
Australia

1. INTRODUCTION

The scene

In this lecture I will discuss some practical and statistical aspects of filtering and spectrum estimation based upon a sequence $(X_t : t = 1,\ldots,N)$ of real numbers. Where appropriate we view this N-tuple as a sample from or part of a

- single   rather than multiple (vector)

- discrete   as opposed to continuous

- time   i.e. one-dimensional index

series.

However a good deal of what I say applies - with appropriate changes - to multiple or continuously indexed arrays, and other generalisations. If our series is sampled from a continuous time series, we assume that the times are equally-spaced.

The problem

The main problem addressed in this lecture is to obtain a good estimate of the (power) spectrum of our series (X_t), as well as an idea of the statistical properties of our estimated spectrum (resolution, bias, precision, etc).

I will also make some brief comments about filtering.

Why do we estimate spectra?

We will not attempt to explain here the reasons why many people : statisticians, communication or control theorists, astronomers, oceanographers, meteorologists, aeronautical engineers and others are interested in calculating spectra. References to work in these and other areas can be found in Tukey (1959) and the other works listed in §6 below.

Why practical?

There are two features of our problems which are una-voidable in seeking pen-and-paper or digital computer solutions:

- the finite length N,   and, where appropriate,

- the discrete (equally spaced) nature of our data.

Coping with these is a practical matter. Doing the computations is another practical matter. We mention the FFT in this context.

Why statistical?

With every piece of data we have to ask: Is it fixed (trend, signal)? Is it random (error, noise)? Is it a mix-ture (sum?) of each?   Of course these questions are related to the questions we wish to ask of the data:

- Are we interested in the unique characteristics of our data?

- Are we interested in our data as a representative of some class of similar data?

● Both of the above?

Perhaps surprisingly, the general techniques for the analysis of a series such as our (X_t) are remarkably similar whichever view we adopt, but of course key differences do emerge. We would not normally attach confidence limits or variances to features of the estimated spectrum of a transient signal, but we can look for periodicities in our data under all circumstances, although the statistical tests for them requires us to make a clear choice.

Mathematically, the notions of limit may differ, as may the meaning of an expression such as mean-squared error, but a clear link between the views comes when

a.   we assume that our series can be extended to $-\infty$ and $+\infty$; and

b.   we make certain regularity assumptions.

The key notion here is ergodicity;   see Brillinger (1975, §2.11) for some further information.

Which spectrum?

Our data $(X_t : t = 1,\ldots,N)$ is assumed to be a finite segment of $(X_t : t = \ldots,-1,0,1,\ldots) = (X_t)$ where (X_t) is a constant mean, covariance stationary random (or stochastic) process. More fully, we assume that

$$\mathbb{E}X_t = \mu \quad \text{is constant}$$

and

$$\mathrm{cov}(X_t,X_{t+k}) = \gamma_k \quad \text{depends only on } k,$$

where $k = \ldots,-1,0,1,\ldots$ is the lag of the covariance. Note that we use $\mathbb{E}$ for the expected value operator and $\mathrm{cov}(U,V)$ to denote the covariance of random variables U and V : $\mathrm{Cov}(U,V) = \mathbb{E}UV - \mathbb{E}U\mathbb{E}V$. Note also that $\gamma_k = \gamma_{-k}$. From now on we will suppose that $\mu = \mathbb{E}X_t = 0$.

Further probabilistic assumptions (strict stationarity, ergodicity, etc.) as often needed for rigorous proofs of some of our results but such matters of rigour will be ignored in this lecture (although we will endeavour to get our formulae correct!).

The <u>spectrum</u> we estimate is

$$f(\omega) = \frac{1}{2\pi} \sum_{-\infty}^{\infty} \gamma_k e^{-i\omega k} = \frac{1}{2\pi} \{\gamma_0 + 2\sum_{1}^{\infty} \gamma_k \cos k\omega\} \qquad (1.1)$$

where we assume $\sum |\gamma_k| < \infty$. Where necessary we write $f_{XX}(\omega)$ to indicate that this is the (auto) spectrum of $X = (X_t)$. As defined this spectrum is a bounded continuous function on $(-\pi,\pi]$, since the <u>summability assumption</u> <u>rules</u> <u>out</u> <u>spectral</u> <u>lines</u>, i.e. discrete components in the spectral measure which would correspond to pure sinusoids. Note that we use angular frequency ω in radians/sec; ordinary frequency is $2\pi\omega$ in sec^{-1}, and I have selected <u>one</u> from the possible choices of $+i$, $-i$, $1/2\pi$, $2\pi i$, $-2\pi i$, $1/\sqrt{2}\pi$, etc. to scale f and ω.

So we are going to estimate $f(\omega)$, a continuous <u>function</u> on $(-\pi,\pi]$, using $X_1,\ldots,X_N$! And know how well we are doing.

2.   EQUALLY SPACED DATA FROM AN UNDERLYING CONTINUOUS RECORD

If our data $(X_t : t = 1,2,\ldots,N)$ originated as equally spaced samples from an underlying continuous process (Y_t) say, with interval $\Delta t : t = \Delta t, 2\Delta t, \ldots, N\Delta t$, then it is likely that our real interest is in the spectrum $f_{YY}(\omega)$ of (Y_t), here defined on $(-\infty,\infty)$ by

$$f_{YY}(\omega) = \frac{1}{2\pi} \int_{-\infty}^{\infty} \gamma_{YY}(\tau) e^{-i\omega\tau} d\tau \qquad (2.1)$$

where $\gamma_{YY}(\tau) = cov(Y_t, Y_{t+\tau})$ depends only on τ, $-\infty < \tau < \infty$, and we assume that $\int_{-\infty}^{\infty} |\gamma_{YY}(\tau)| d\tau < \infty$.

How does f_{XX} related to f_{YY}? This is the well-understood phenomenon of <u>aliasing</u> - see any of the references - the relevant formula being

$$f_{XX}(\omega) = \sum_{k=-\infty}^{\infty} f_{YY}(\omega + \frac{2k\pi}{\Delta t}) , \quad \frac{-\pi}{\Delta t} < \omega \leqslant \frac{\pi}{\Delta t}. \qquad (2.2)$$

The upper limit of the range of the sampled process (X_t) is called the Nyquist (folding) frequency and is

$$\omega_N = \frac{\pi}{\Delta t} \text{ rad sec}^{-1} \text{ (or } \frac{1}{2\Delta t} \text{ sec}^{-1}\text{)}.$$

The aliases of $\omega \varepsilon [0, \frac{\pi}{\Delta t}]$ are $\pm\omega + 2k\omega_N$, $k = \pm 1, \pm 2, \dots$.

The selection of sampling rate is equivalent of selecting the Nyquist frequency and this can only be done with some confidence if there is some knowledge of f_{YY}.

For the remainder of this lecture we suppose that $\Delta t = 1$ and ignore any aliasing problems, referring to the books listed for further assistance.

3. ESTIMATION OF SPECTRA

There are many different reasons why one might wish to estimate a spectrum and these different reasons frequently imply the desire to emphasise different features of the spectrum, which in turn imply the need to use different methods of estimation. Thus one method might be appropriate to give a quick global estimate of the spectrum, whilst another is needed to permit the close examination of a local region in which there is not much power; one might produce smoother estimates than another, whilst one might be better than another at resolving close peaks. It should be remembered that the spectral ordinates generally range over a number of orders of magnitude and hence are best examined on a log scale.

There are two main methods of spectral estimation: the non-parametric, generally based upon quadratic functions of the data, which is (roughly) equivalent to smoothing the periodogram; and the maximum likelihood (ML) and related methods including the auto-regressive (AR) or more recent autoregressive-moving average (ARMA) methods, which are based upon estimating the parameters of a finitely parametrized spectrum, and which return an estimate which is a rational function whose coefficients are highly non-linear functions of the data. Note that the maximum entropy (ME) method falls into this second class. In this lecture we will only discuss

periodogram methods, referring to the literature for details
of the others: Kay and Marple (1981), Thompson (1982).

One piece of advice (which may not be all that easy to
follow) is: always try to give two (or more) estimates of
any spectrum, one (possibly others) non-parametric (the others with a different amount of smoothing) and one AR (or
ARMA); then put both (or all) on the same graph. An additional aid would be an estimate of the one-step prediction
error variance in each case, something which is well known
with the AR or ARMA approach, and which can be obtained in
the non-parametric approach via a famous formula of Szegö,
see Hannan and Nicholls (1977). Kay and Marple (1981) have
an interesting figure (p.1409) illustrating 12 different
estimtion methods applied to the same data.

4. THE PERIODOGRAM

The discrete Fourier transform (DFT) of $(X_t: t=1,\ldots,N)$
is

$$Z_k^{(N)} = \frac{1}{\sqrt{2\pi N}} \sum_1^N X_t e^{-i\omega_k t} \tag{4.1}$$

where $\omega_k = \dfrac{2\pi k}{N}$, defined for $k = 1,\ldots,N$. Their squared
moduli

$$I_k^{(N)} = \frac{1}{2\pi N} \left| \sum_1^N X_t e^{-i\omega_k t} \right|^2 \tag{4.2}$$

are known as the periodogram ordinates. [We could define
$Z_{(N)}^{(N)}(\omega)$ for all values of ω in $(-\pi,\pi]$, and similarly for
$I^{(N)}(\omega)$, but such functions are fully determined by the
values at the frequencies $\{\omega_k\}$.] An important re-expression
of $I_k^{(N)}$ is as

$$I_k^{(N)} = \frac{1}{2\pi} \left\{ c_0 + 2 \sum_1^{N-1} c_r \cos r\omega_k \right\} \tag{4.3}$$

where $c_r = \dfrac{1}{N} \sum_{r+1}^N X_t X_{t-r}$ for $r \geqslant 0$ and $c_{-r} = c_r$, $r < 0$.

The similarity of (4.3) with the definition of $f = f_{xx}$ is unmistakeable; indeed we have the formulae:

$$\mathbb{E}I_k^{(N)} \approx f(\omega_k) \tag{4.4a}$$

$$\text{Var } I_K^{(N)} \approx f(\omega_k)^2 \qquad k \neq 0, \tfrac{1}{2}N \text{ (double in these cases)} \tag{4.4b}$$

$$\text{Cov}(I_k^{(N)}, I_\ell^{(N)}) \approx 0, \qquad k \neq \ell \tag{4.4c}$$

and these formulae are the starting points for most non-parametric estimates of the spectrum. We examine each of them in turn, bearing in mind the fundamental statistical relation:

$$\text{MEAN SQUARED ERROR } = \text{ VARIANCE } + \text{ (BIAS)}^2$$

where VARIANCE and STANDARD DEVIATION = $\{\text{VARIANCE}\}^{1/2}$ measure precision and BIAS measures accuracy.

(i) Bias (resolution) of an estimated spectrum; all related to (4.4a) which we rewrite exactly as

$$\mathbb{E}I_k^{(N)} = \int_{-\pi}^{\pi} |\phi^{(N)}(\omega - \omega_k)|^2 f(\omega)d\omega \tag{4.4a}$$

where

$$\phi^{(N)}(\omega) = \frac{1}{\sqrt{2\pi N}} \sum_1^N e^{i\omega t}$$

and

$$|\phi^{(N)}(\omega)|^2 = \frac{\sin^2 \tfrac{1}{2}N\omega}{2\pi N \sin^2 \tfrac{1}{2}\omega} .$$

This is the Fejer kernel which approximates the delta functions as $N\uparrow\infty$, but it decreases rather too slowly, leaving the possibility that a (relatively large) value of f at a point ω some distance away from ω_k might bias our estimate of the (relatively small) value of $f(\omega_k)$. This phenomenon is known as leakage.

Correcting for leakage – at a price – can be done by the use of a data taper (fader, window) $a_t^{(N)}$ as follows, cf. Hannan (1970, p.266):

Replace $Z_k^{(N)}$ by

$$\frac{1}{\sqrt{2\pi N}} \sum_1^N a_t^{(N)} X_t e^{-i\omega_k t} \tag{4.5}$$

and, instead of the raw periodogram $I_k^{(N)} = |Z_k^{(N)}|^2$, use

$$g(\omega_k) = \frac{1}{2\pi N} \left| \sum_1^N a_t^{(N)} X_t e^{-i\omega_k t} \right|^2 \tag{4.6}$$

for which we have the formula

$$Eg(\omega_k) = \int_{-\pi}^\pi |A^{(N)}(\omega-\omega_k)|^2 f(\omega) d\omega \tag{4.7}$$

where $A^{(N)}(\omega) = \frac{1}{\sqrt{2\pi N}} \sum_t a_t^{(N)} e^{i\omega t}$. How do we choose the taper

to reduce the bias in (4.7)? Typically we would have $a_t^{(N)} = u(\frac{t}{N})$ or $u(\frac{t-\frac{1}{2}}{N})$ for some continuous function u which is zero off $[0,1]$, and in such cases

$$N^{-\frac{1}{2}} A^{(N)}(\frac{\omega}{N}) = \frac{1}{\sqrt{2\pi}} \frac{1}{N} \sum_1^N u(\frac{t}{N}) e^{i\omega t/N} \to \frac{1}{\sqrt{2\pi}} \int_0^1 u(x) e^{i\omega x} dx = \hat{u}(\omega)$$

as $N \to \infty$. Rewriting (4.7) as

$$Eg(\omega_k) = \int N^{-1} |A^{(N)}(\frac{\omega}{N})|^2 f(\omega_k - \frac{\omega}{N}) d\omega \tag{4.8}$$

we can see that our aim in choosing u to make the bias in (4.8) as small as possible is realised by asking that $|\hat{u}(\omega)|^2$ decrease to zero as fast as possible as $|\omega|$ increases. Now the higher the order of contact of u with the horizontal axis at $0,1$ the faster $|\hat{u}(\omega)|^2$ decreases as $|\omega|$ increases, i.e. the faster $N^{-1}|A^{(N)}(\frac{\omega}{N})|^2$ decreases as $|\omega|$ increases.

Our initial (default) taper corresponds to $u_D(x) \equiv 1$ on $[0,1]$ (and $= 0$ off $[0,1]$) and has

$$N^{-1} |A^{(N)}(\tfrac{\omega}{N})|^2 \sim |u_D(\omega)|^2 \sim \omega^{-2}$$

as $|\omega|$ increases, whereas $u_H(x) = \frac{1}{2}(1-\cos 2\pi x)$

has $|u_H(\omega)|^2 \sim \omega^{-6}$ as $|\omega|$ increases. The taper based upon u_H defines what is known as hanning.

It is common to use a combination of these two such as the cosine bell

$$a_t^{(N)} = \begin{cases} \frac{1}{2}[1-\cos(\frac{\pi(t-m)}{m}] & 1 \leqslant t \leqslant m \\ 1 & m < t \leqslant N-m \\ \frac{1}{2}[1-\cos(\frac{\pi(N-t+m)}{m})] & N-m < t \leqslant N \end{cases}$$

where m is suitably chosen, e.g. so that the proportion $P = 2m/N$ of data tapered is 10% or 20%. It can be shown that such tapers give a definite reduction in leakage.

What do we lose by tapering? The answer is: not much ! If $g(\omega_k)$ is defined by (4.6) then

$$\mathbb{E}g(\omega_k) \approx U_2 f(\omega_k)$$

where $U_2 = \frac{1}{N} \sum_1^N \{a_t^{(N)}\}^2 \approx \int_0^1 u(x)^2 dx$, whilst

$$\text{Var } g(\omega_k) \approx U_4 f(\omega_k)^2 \qquad k \neq 0, \tfrac{1}{2}N ,$$

where $U_4 = \frac{1}{N} \sum_1^N \{a_t^{(N)}\}^4 \approx \int_0^1 u(x)^4 dx$. The variance of

the logarithm of a spectrum estimate is multiplied by U_4/U_2^2 and from Bloomfield (1976, p.194) we find that the split cosine bell taper with $P = 2m/N$ taking the values 0.1 and 0.2, the ratio U_4/U_2^2 has the values 1.055 and 1.116 respectively. A small price to pay for protection against leakage.

(ii) Variance (stability, precision) of an estimated spectrum

From (4.4b) we see that the variance of the raw periodogram ordinate at ω_k is (approximately) constant; it does
not decrease as $N \to \infty$. We need to pool adjacent periodogram ordinates - assuming a certain smoothness of the underlying spectrum - to get an estimate with some stability.
This works because of (4.4c): the adjacent (or any other)
ordinates are approximately uncorrelated.

Care is needed! In averaging adjacent periodogram ordinates we run the risk of introducing bias - leakage - if
widely differing values of the spectrum occur over short frequency ranges.

[The easiest spectrum to estimate is a flat one, for
then leakage is not a problem. This is why Blackman and
Tukey (1958) recommended prewhitening: filter to flatten
your spectrum, estimate the result and then correct for the
filter. Still good advice today.]

And so, just as we needed to avoid sharp corners in our
data tapering, so we need to in forming spectral averages

$$g(\omega_k) = \sum_n g_n I^{(N)}_{k-n} \tag{4.9}$$

where $\{g_n\}$ is a set of weights. For example, leakage can
occur with

$$g_n = \begin{cases} \dfrac{1}{2m+1} & n = -m, \ldots, m \\ 0 & \text{otherwise} \ ; \end{cases} \tag{4.10}$$

for in this case we are simply averaging any given periodogram ordinate with the m on either side. Expression for
the actual bias involve the nature of the window and usually
the second derivative $f''(\omega)$ of the spectrum, see Jenkins
and Watts (1968, §6.3), and in much practice the bias may not
be a problem.

What gains do we make using spectral averages? The discussion which follows is based on §8.5 of Bloomfield (1976).
For an estimate $g(\omega)$ of the form (4.9) we can easily obtain
the approximate relations

$$\mathbb{E}g(\omega) \approx f(\omega)\Sigma \ g_n$$
$$n$$

$$\text{Var } g(\omega) \approx f(\omega)^2 \ \Sigma \ g_n^2$$
$$n$$

and of course averaging will introduce a degree of correlation between our estimates unless it is done on disjoint blocks of frequencies. The improvement comes from having $\Sigma_n \ g_n^2 < 1$; indeed it is $1/(2m+1)$ with example (4.10).

To carry out further analysis let us suppose that our spectral averages (4.9) have <u>spectral weights</u> $\{g_n\}$ of the form $g_n = 2\pi/N^{-1}W(\omega_n)$ in terms of some smooth function $W(\omega)$ called a <u>spectral window</u>. In such cases (4.9) is approximately

$$g(\omega) \approx \int_{-\pi}^{\pi} \ W(\lambda)I^{(N)}(\omega-\lambda)d\lambda \qquad (4.11)$$

where $I^{(N)}(\omega) = (2\pi N)^{-1} \ |\sum_{1}^{N} X_t e^{-i\omega t}|^2$. For estimates of the form (4.11)

$$\text{Var } g(\omega) \approx \frac{2\pi}{N} \ f(\omega)^2 \ \int_{-\pi}^{\pi} \ W(\lambda)^2 d\lambda \ .$$

An alternative class of estimators of $f(\omega_k)$ are the weighted autocovariance estimators, which arise by weighting the $\{c_r\}$ in (4.3) and thus take the form

$$g(\omega_k) = \frac{1}{2\pi} \ \{w_0 c_0 + 2 \ \sum_{1}^{N-1} \ w_r c_r \cos r\omega_k\} \ . \qquad (4.12)$$

By the convolution theorem this is just (4.11) again where

$$W(\lambda) = \frac{1}{2\pi} \ \Sigma \ w_r e^{-ir\lambda}$$
$$r$$

and we see that smoothing the periodogram or weighting (tapering) the auto-covariances gives the same class of spectral estimators. If we have $w_r = w(r/M)$ where $w(x) = 0$ if $|x| \geq 1$, then estimates (4.11) or (4.12) have

$$\text{Var } g(\omega_k) \approx \frac{M}{N} f(\omega)^2 \int w(x)^2 dx$$

and in this form we see the role of the proportion M/N of non-zero lag weights.

The literature contains a lot of examples of windows which have been designed to minimise leakage at the same time as increasing stability, see e.g. Brillinger (1975, p.55), and of course there are theorems guaranteeing consistency as $M,N \to \infty$ in the preceding formula. The key concept in all discussions of window is that of (equivalent) bandwidth, see e.g. Koopmans (1974, p.277).

To obtain the variance of a smoothed periodogram (or weighted covariance) estimator of the spectrum when data tapering has also taken place, simply combine the expressions given in sections (i) and (ii). Perhaps the nicest formula expressing both these effects is that given as Exercise 8.9 of Bloomfield (1976, p.199):

- for a Gaussian white noise series (X_t) with variance σ^2 tapered with $\{a_k\}$, a spectrum estimate with lag weights $\{w_r\}$

$$g(\omega) = \frac{1}{2\pi N} \sum_s \sum_t a_s a_t X_s X_t w_{s-t} e^{-i\omega(s-t)}$$

has, when $a_k = u(\frac{k-\frac{1}{2}}{N})$, approximate mean and variance

$$\mathbb{E}g(\omega) \approx \frac{\sigma^2}{2\pi} \int_0^1 u(x)^2 dx \quad ,$$

$$\text{Var}g(\omega) \approx \frac{\sigma^4}{2\pi N} \sum_r w_r^2 \int_0^1 u(x)^4 dx \quad , \quad \omega \neq 0, \pi \quad .$$

(iii) Confidence intervals for the true spectrum $f(\omega)$ or, more importantly, $\log f(\omega)$ can be obtained by assuming approximate joint normality of the estimates if N is large and the ratio

$$\alpha^2 = \frac{\text{varg}(\omega)}{f(\omega)^2}$$

is not too large. For example, a 95% approximate confidence

interval for log $f(\omega)$ (natural logs) would be just

$$\log g(\omega) \pm 1.96\alpha \;.$$

A more accurate method uses a chi-squared approximation (χ^2_ν) with equivalent degrees of freedom $\nu = 2/\alpha^2$. More precisely, it is assumed that with this ν, we have

$$\frac{\nu g(\omega)}{f(\omega)} \sim \chi^2_\nu$$

and so can obtain the 95% confidence interval

$$\nu g(\omega)/\chi^2_\nu(0.975) \leqslant f(\omega) \leqslant \nu g(\omega)/\chi^2_\nu(0.025),$$

where $\chi^2_\nu(0.975)$ and $\chi^2_\nu(0.025)$ are the 2.5% and 97.5% points of a chi-squared distribution with ν degrees of freedom.

A simultaneous confidence band can also be constructed for large samples, Bloomfield (1976, p.197), and this would be wider by a factor of about $\log m$ where $m = 2\pi/\text{bandwidth}$. We refer to Brillinger (1975, §3.3) for a good discussion of the notion of bandwidth of a window.

(iv) <u>Doing the calculations</u>

Since the late 1960s spectrum estimation has been almost universally carred out according to the following sequence:

1. pad out the data with zeros to obtain a series of length $N^\prime \geqslant N$ where $N^\prime$ is a power of 2;

2. taper the data in some way;

3. use the fast Fourier transform (FFT) to get the discrete Fourier transform of the (padded and tapered) data;

4. form the sum of squares of the real and imaginary parts of (half of) the Fourier transform giving a raw periodogram;

5. smooth the raw periodogram in some way.

I have already given some information about data tapers in (ii). You can read about the FFT in <u>any</u> (recent) book or

in the special issues of June 1967 and June 1969 of IEEE
Trans. Audio Electroacoust. AU-15, AU-17 resp.   There is a
lot to be said on this topic.

The question (1) of padding out the data with zeros to a
length which is a power of 2 deserves a little more comment,
see §A of Tukey (1980). In that reference Tukey explains a
phenomenon which he calls separation aliasing which is a
consequence of squaring the discrete Fourier transform. As
with the other kind of aliasing, one can try to understand it
and one's own problem enough to be able to decide whether it
can be ignored or whether steps need to be taken to avoid it.
If data of length N are padded with at least N zeros,
then the problem is eliminated.

Even if one only wants the estimated autocovariances
$\{c_r\}$, the mean lagged products, the sequence (1), (2) (3) and
(4) above, followed by an inverse FFT, is – for not too large
N – an efficient way to proceed.

(v)   Final comment

The foregoing is a summary of the conventional wisdom
following Blackman and Tukey (1958) and accordingly, is
rather old-fashioned. The 1970s saw an explosion of activity
in the area of spectrum estimation which is summarised rather
well by the paper of Kay and Marple (1981) and the September
1982 issue of the IEEE Proceedings, Volume 70. Much of this
activity derived from the desire to get reasonable resolution
with quite short series. However other methods such as
Thompson's (1982), using prolate spheroidal wave functions,
are excellent alternatives to traditional methods.

5.   FILTERING

Much of the basic theory of linear time invariant
filters concerns the construction and characteristics of
ideal filters, i.e. filters which are optimal in some sense,
such as passing a given band of frequencies. Once again
practicalities intrude on us, for

   a. we will only apply our filters to a finite stretch of
       data, and

b. ideal filters generally have infinitely many non-zero
    weights and can become distinctly non-ideal when trun-
    cated.

Here we assume that all filters can be realised as convol-
tuion filters with a given set of weights:

$$X_t \to \Sigma_k \, a_k X_{t-k} = Y_t.$$

and in general our interest is focussed upon filters with
only finitely many non-zero weights, although an important
class called <u>recursive</u> filters do not have this property. In
general the <u>transfer</u> function of the convolution filter with
weights $\{a_k\}$ is

$$A(\omega) = \Sigma_k \, a_k e^{-i\omega k}$$

with $|A(\omega)|$ and $\arg A(\omega)$ being known as the gain and
phase shift functions, respectively, of the filter. Note
that our restriction to linear filters excludes consideration
of certain other popular procedures such as median filters.

(i) Effect of finite data length

One of the major concerns with the application of a convolution filter is: What do we do at the ends? Behave as if
the data were zero for $t = 0, -1, \ldots$ and $t = N+1, N+2, \ldots$?
Copy the end values? Match up $t = N+1$ with $t = 1$? Omit a
stretch at each end? See Bloomfield (1976, pp.123-5) for a
discussion of these points.

Koopmans (1974, §6.3) gives a short discussion of the
differences arising when a filter with weights $\{a_k\}$ is
applied to a series $(X_t : t=1, \ldots, N)$ extended beyond its range
by zeros. He shows that if U_t is the outcome of this
analysis, and Y_t is what would result if we used the whole
sequence $(X_t : t = \ldots, -1, 0, 1, \ldots)$, then

$$2\pi m S \leqslant \mathbb{E}(U_t - Y_t)^2 \leqslant 2\pi M S \qquad (5.1)$$

where $S = \Sigma\{a_k^2 : k < t-N, \; k > t-1\}$ and we have assumed that the
spectral density $f(\omega)$ of X satisfies $0 < m \leqslant f(\omega) \leqslant M$.
Since we have

$$S = \sum_{-\infty}^{t-N-1} a_k^2 + \sum_{t}^{\infty} a_k^2$$

it is clear that the magnitude of the difference in (5.1) depends not only on the spectrum of X but on the rate of decrease of the weights $\{a_k\}$. Ideal filters such as those having transfer functions of the boxcar type (linear combinations of Heaviside functions) have slowly decreasing weights (Fourier coefficients) and thus less than ideal performance, particularly as we might prefer to omit values near the ends of our data, but do not want to omit too many data points. Just as we found in the discussion of data tapering, filters with smoother transfer functions have weights which decrease faster and so are preferable in practice. Indeed we can taper filter weights as well, see Koopmans (1976, pp.187-9).

(ii) <u>Linear filtering by the FFT</u>

Because of the cheapness of FFTing data in comparison to calculating convolutions – see also the comment about calculating mean lagged products in 4(iv) above – it is important to understand the error involved in carrying out the sequence

(1)   FFT the data;

(2)   Multiply the result of (1) by the transfer function of your filter;

(3)   unFFT the result of (2).

If we only have the filter weights we need to interpolate the step

(1½)   FFT the filter weights to get (an approximation to) the transfer function of your filter.

Once more we refer to Koopmans (1976,§6.3 and A6.3) for the discussion whose results we summarize. If a filter with weights $\{a_k\}$ has (exactly) known transfer function $A(\lambda)$ and we implement it by the sequence (1), (2), and (3) above, then the actual result V_t differs from the theoretically expected result Y_t in mean square as follows:

$$2\pi m S_{N,t} \leqslant E(V_t - Y_t)^2 \leqslant 2\pi M S_{N,t}$$

where, as in (i) above, we have $m \leqslant f(\omega) \leqslant M$ for the spectrum $f = f_{XX}$ of X and

$$S_{N,t} = \sum_{-\infty}^{t-N-1} a_k^2 + \sum_{t}^{\infty} a_k^2 + \sum_{t-N}^{t-1} (a_k^2 - \tilde{a}_k)^2$$

where

$$\tilde{a}_k = \sum_{p=-\infty}^{\infty} a_{k+pN}.$$

(iii) Filtering and decimation

Our final and important point again concerns the need to low-pass filter data before resampling at a multiple of the original sampling interval, a process known as decimation.

If data samples at an interval Δt are to be subsampled to an interval $(\Delta t)' = s\Delta t$, reducing N data points to $N' = N/s$, then the Nyquist frequency ω_N of the new series will be smaller than that of the original series, indeed

$$\omega_N' = \frac{\omega_N}{s}.$$

In order to avoid aliasing it will be necessary to pass the original data through a filter with a cut-off frequency $\omega_0 \leqslant \omega_N'$. In practice some smooth approximation to the ideal low-pass filter with cut-off ω_N' will be used. Here as in data tapering and the smoothing of periodograms, we will want to avoid leakage, i.e. large side lobes in the transfer function of our low-pass filter.

There are many books on the design of digital filters!

REFERENCES

1. R.B. Blackman and J.W. Tukey, "The measurement of Power Spectra," Dover, New York (1958).
2. P. Bloomfield, "Fourier Analysis of Time Series: an Introduction," Wiley, New York (1976).
3. D.R. Brillinger, "Time Series: data analysis and theory," Holt, Rinehart & Winston, Inc., New York (1975).
4. E.J. Hannan, "Multiple Time Series," Wiley, New York (1970).

5. E.J. Hannan and D.F. Nicholls, The estimation of the prediction error variance, J. Amer. Statist. Assoc., 72, (1977), 834–840.

6. IEEE Proceedings (September 1982) Volume 70, Number 9 Spectral Estimation, Thirteen papers, 885–1124.

7. G.M. Jenkins and D.C. Watts, "Spectral Analysis and its Applications," Holden Day, San Francisco (1968).

8. S.M. Kay and S.L. Marple Jr., Spectrum analysis – a modern perspective, Proc. IEEE, 69 (1981), 1380–1419. [278 references]

9. L.H. Koopmans, "The Spectral Analysis of Time Series," Academic Press, New York (1974).

10. M.B. Priestley, "Spectral Analysis and Time Series," Volumes 1,2, Academic Press, London (1981).

11. D.J. Thompson, Spectrum estimation and harmonic analysis, Proc. IEEE, 70 (1982), 1055–1096. [362 references]

12. Probability and Statistics, The Harald Cramer Volume, ed. U. Grenander. Almqvist & Wiksell, Stockholm. pp.300–330.

13. J.W. Tukey, Can we predict where "Time Series" should go next? Directions in Time Series, Proceedings of the IMS Special Topic Meeting on Time Series Analysis, Iowa State University, May 1978. Ed. D.R. Brillinger and G.C. Tiao. pp.1–31.

PART 2

FOURIER TECHNIQUES AND APPLICATIONS

TWO-DIMENSIONAL PHASE RESTORATION

R.H.T. Bates and W.R. Fright

Electrical and Electronic Engineering Department
University of Canterbury
Christchurch, New Zealand

ABSTRACT

   The constraints laid on the phase of a Fourier transform
by its intensity are reviewed in the contexts of well known
phase problems. The considerable differences between phase
problems involving one-dimensional and multi-dimensional
images, and finite-sized (as arise in astronomy, for
instance) and periodic (as occur in crystallography) images,
are explained. The crucial importance, for uniqueness questions, of the concept of the image-form (and also its most
compact manifestation) is emphasised, as is the almost always
unique connection between the image-form of a positive
multi-dimensional image and the intensity of its Fourier
transform. The current status of phase recovery algorithms,
as regards Fourier transforms of finite-sized images, is
assessed. The necessity for composite algorithms, incorporating simple but powerful constructions, is pleaded and
reinforced by computational examples illustrating our previously reported defogging routine and a new procedure called
fringe magnification.

1. INTRODUCTION

   Suppose one knows that two functions, $f(x,y)$ and $F(u,v)$,
form a Fourier transform pair, meaning that the one is the
Fourier transform of the other, and the other is the Fourier

transform of the one, which connection we identify by

$$f(x,y) \leftrightarrow F(u,v) \tag{1}$$

It is appropriate to call $f(x,y)$ the image and $F(u,v)$ the
visibility, with (x,y) and (u,v) being the Cartesian coordinates of arbitrary points in two-dimensional image space and
two-dimensional Fourier space respectively. Both the image
and the visibility are in general complex quantities. We
find it convenient to adopt the notation

$$Z = |Z| \exp(i \text{ phase}\{Z\}) \tag{2}$$

for identifying the magnitude (or modulus, or absolute value)
and phase - i.e. $|Z|$ and phase$\{Z\}$ respectively - of any com-
plex quantity Z, whose complex conjugate and intensity we
write as Z^* and $|Z|^2$ respectively.

Suppose further that one is presented with merely the
visibility magnitude $|F(u,v)|$. What can be inferred about
phase$\{F(u,v)\}$ in the absence of extra information? In addition, can we come to sharper conclusions when the image is
subject to particular constraints? Such questions are
aspects of what we call the Fourier phase problem.

Two images: $g(x,y)$ on the left and $h(x,y)$ on the right.

Mathematically educated people who come upon this problem for the first time tend to be surprised that it can have
much content. After all, why should one not attach "any old"
phase to the given magnitude? Such impressions are strongly

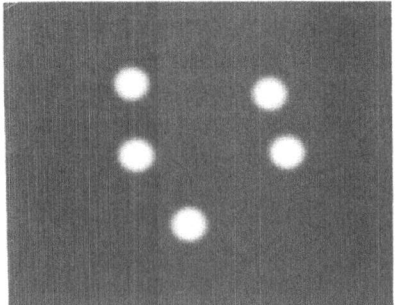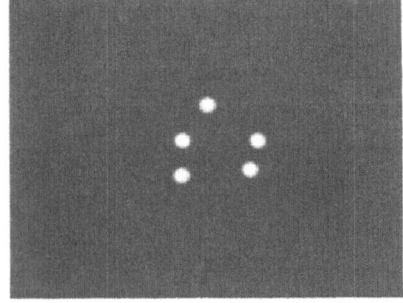

Figure 1

reinforced by the well known "dominance" of phase over magnitude (illustrated in §2 below). It is nevertheless true that the visibility phase is constrained by the visibility magnitude in scientifically important ways. If this was not so, we would of course have been spared our present labours!

After posing the two-dimensional phase problem in a precise fashion in §3, we give a brief history of phase problems in §4. Various preliminaries are collected in §5 and §6, and we point out in the latter why phase problems involving images of finite size (which are the concern in this paper) are quite different from crystallographic phase problems. Further preliminaries are introduced in §7. The next three section, §8 through §10, are devoted to explaining why and in what respect multi-dimensional phase problems involving finite-sized images can be expected to possess unique solutions. The details of the argument are probably more plausible than rigorous (but that is the way of the scientific world; if we were always to attend upon logical certainty, very little would get done!). In the remaining sections, §11 through §14, we describe practical phase recovery algorithms, illustrating them with computational examples, and assess future possibilities.

2. DOMINANCE OF PHASE OVER MAGNITUDE

Figure 1 shows a pair of images, denoted by $g(x,y)$ and $h(x,y)$, each consisting of 5 isolated "blobs" of brightness. Note that the blobs in $g(x,y)$ are larger than those in $h(x,y)$. We have computed the respective visibilities, $G(u,v)$ and $H(u,v)$, of these two images. The second pair of images $g(x,y|h)$ and $h(x,y|g)$, shown in Figure 2, are defined respectively by

$$g(x,y|h) \leftrightarrow |G(u,v)| \exp(i \text{ phase}\{H(u,v)\}) \qquad (3)$$

and

$$h(x,y|g) \leftrightarrow |H(u,v)| \exp(i \text{ phase}\{G(u,v)\}. \qquad (4)$$

Consider the brightest blobs in the two images shown in Figure 2. There are five such blobs in each image. The juxtaposition of the brightest blobs in $g(x,y|h)$ is almost the same as that in $h(x,y)$, and the brightest blobs in $h(x,y|g)$

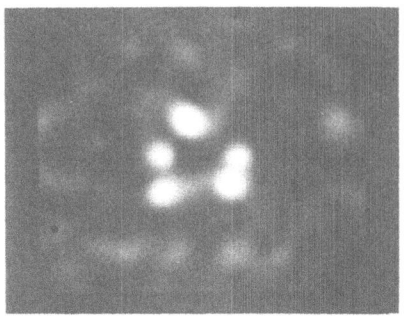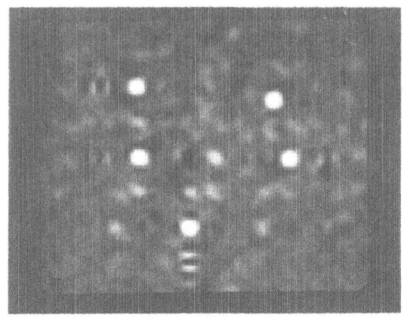

Figure 2

Two more images: $g(x,y|h)$ on the left and $h(x,y|g)$ on the right.

are positioned similarly to those in $g(x,y)$. This phase-dominance caused some heart-searching amongst X-ray crystallographers (cf. Ramachandran and Srinivasan 1970) when it was first widely recognised. It seemed to confirm the fears of those who thought that phase problems might be essentially vacuous.

The truth of the matter is, while the visibility phase is unquestionably dominant, it is nevertheless palpably constrained by the visibility magnitude, as we demonstrate in what follows. The implicit constraints are far more powerful than those apparent in Figure 2, the strongest of which is that the sizes of the brightest blobs in $g(x,y|h)$ and $h(x,y|g)$ correspond to those in $g(x,y)$ and $h(x,y)$ respectively. This is merely due to the effective range of the spatial frequencies possessed by $g(x,y|h)$ being set by $g(x,y)$, and similarly for $h(x,y|g)$ and $h(x,y)$, as inspection of (3) and (4) shows. The more powerful constraints are áctually hidden in the diffuse background brightnesses apparent in Figure 2. The only way to remove these backgrounds is to associate the "correct" phases with each of the magnitudes $|G(u,v)|$ and $|H(u,v)|$. The really extraordinary thing is that, although this is a delicate business in some senses, it is remarkably robust in others.

3. THE FOURIER PHASE PROBLEM

We pose the Fourier phase problem as: given $|F(u,v)|$, recover the image-form of $f(x,y)$.

For arbitrary constant ξ_1, ξ_2, η_1 and η_2, we say that $f(x - \xi_1, y - \eta_1)$ and $f^*(-x + \xi_2, -y + \eta_2)$ have the same image-form as $f(x,y)$. The motivation for this definition is that an image remains recognisable when it is moved or is viewed in a mirror. Even though the mirror in question has the peculiar property of conjugating the image (!), it seems reasonable to argue that $f(x,y)$ and $f^*(-x,-y)$ look the same because their magnitudes are equal at points reflected in the origin of image space. It is worth remembering that the image is positive (by which is here meant that it is real and non-negative) in many of the more important practical applications.

The need for introducing the image-form arises because it is impossible to distinguish $F(u,v)$ from

$$(F(u,v)\exp(i[\xi_1 u + \eta_1 v]))$$

or from

$$(F^*(u,v) \exp(i [\xi_2 u + \eta_2 v])$$

when only $|F(u,v)|$ is given.

We have already made an oblique reference (in §2) to a certain class of phase problem; that which occurs in X-ray crystallography. It is important to emphasise that the crystallographic phase problem differs in one essential particular from the problem with which this paper is concerned. We discuss this in some detail in §6 below.

4. PHASE PROBLEMS REVIEWED

It is probably fair to say that the only Fourier phase problems, for which rigorous non-trivial mathematical results exist, are one-dimensional. In such problems, we are asked to infer as much as possible about phase$\{F(u)\}$, given $|F(u)|$ on the understanding that $f(x) \leftrightarrow F(u)$, where x and u are scalar variables, implying that image space and Fourier space are one-dimensional.

Interesting results only arise when the image is of finite size (or of finite extent, or has compact support). It
must be emphasised that this introduces no practical con-
straint whatsoever because the sizes of all images which
occur in scientific and technical contexts are necessarily
finite. It transpires that a set $\{f(x|j); j=1,2,...,J\}$ of
image-forms exists for which

$$f(x|j) \leftrightarrow |F(u)| \exp(i \ phase\{\Phi(u|j)\}) \tag{5}$$

All of the $f(x|j)$ occupy the same interval of the x-axis –
i.e. they are all identically zero outside this interval, and
they all have finite value near the extremities of this
interval (they almost always have finite value throughout
this interval, except perhaps at discrete points). The
phases $\Phi(u|j)$ differ appreciably, implying (see §2 above)
that all the $f(x|j)$ are significantly different from each
other. The value of J is, theoretically, almost always ∞.
Even though practical considerations force J to be effectively finite, it usually remains large. All of this is
clearly explained and made precise by Taylor (1981), who
references the relevant literature.

The rampant non-uniqueness of one-dimensional phase
problems can be attributed to the fundamental theorem of
algebra. It happens that F(u) is necessarily entire when
f(x) is of finite length. The aforementioned theorem ensures
that F(u) can always be factored into (an infinite number of)
terms such as $(u-\zeta(k))$, where k is an integer index and $\zeta(k)$
is in general complex. Given merely $|F(u)|$, we cannot prefer
$\zeta(k)$ to $\zeta^*(k)$. If all of the $\zeta(k)$ are exchanged for their
complex conjugates, the sign of the phase is reversed, so
that the image-form is unaltered. Consequently, if K of the
$\zeta(k)$ have non-zero imaginary parts,

$$J = 2^{K-1} \tag{6}$$

It was therefore surprising when isolated results began
appearing some ten years ago (cf. Napier and Bates 1974; Gull
and Daniell 1978) suggesting that the "degree of non-
uniqueness" is somehow much less for two-dimensional than for
one-dimensional problems. Most of us thought these must be
"special cases". However, the number of apparently unique
examples multiplied until it was clear that some essential
change takes place in passing from one to two dimensions.
Fienup's insight and perseverance has been the main cause of

the alteration of viewpoint (cf. §7, Bates 1982b). We refer
the reader especially to Fienup's (1982) paper which compares
the various algorithms he has pioneered.

Bruck and Sodin (1979) provide the crucial intuition,
which can be paraphrased by the observation that there is no
fundamental theorem of algebra in more than one dimension.
Even though F(u,v) is necessarily an entire function of both
u and v, when f(x,y) is of finite size, it is almost always
prime, meaning that it is only factorisable in very special
cases. The previous source of "non-uniqueness" therefore
evaporates. The upshot is that there is almost always only
one image-form covering the smallest area of image space compatible with a given visibility intensity. When the image is
known a priori to be positive, its image-form cannot cover an
area larger than the aforementioned smallest one (Bates 1984;
Bates and Fright 1984). These uniqueness conditions seem to
apply with growing force as the dimensionality increases.

While doubts continue to be raised (cf. Huiser and van
Toorn 1980; Kiedron 1981; Fiddy, Brames and Dainty 1983), the
available evidence (cf. Fright and Bates 1982; Bates and
Fright 1983, 1984; Bates, Fright and Norton 1984; Fienup
1984) increasingly suggests that "uniqueness" is guaranteed
except in contrived situations. Much needs to be done, of
course, before visibility phases can be routinely recovered
from their magnitudes under conditions pertaining in important practical situations. We hope that the following
disquisition will convince the reader that this goal is now
at least in sight.

5.   NECESSARY PRELIMINARIES

It is rare in practical applications for $|F(u,v)|$ to be
recorded as a continuous function of u and v. We are usually
presented with samples of the visibility magnitude. Before
considering how closely spaced these samples must be, it is
convenient to introduce notation which characterises an image
of finite size. We denote by Bf = Bf(x,y) the smallest rectangle (we actually take it to be square, for convenience,
which loses no generality because the distance units in, say,
the y-direction can be altered to make the measured lengths
of all sides of the rectangle equal) which just encloses

f(x,y). We call Bf the image box and normalise its size by

$$Bf(x,y) = 1 \quad \text{for} \quad 0 < (|x|,|y|) < 1/2N$$
$$= 0 \quad \text{for} \quad 1/2N < (|x|,|y|) < \infty \tag{7}$$

where the positive integer N is introduced for later convenience. Note that the length of each side of the image box is
1/N.

The sampling theorem (cf. Bracewell 1978) ensures that
F(u,v) is completely determined by its samples F(mN,nN), for
all values of the integers m and n for which the visibility
magnitude is significant (this means, in any practical case,
where the magnitude exceeds the "noise level"). The essence
of the matter is that the sample spacing is N in both the u
and v directions. This is only useful when the phases as
well as the magnitudes of the samples are known.

It is a theorem (cf. Bracewell 1978) that

$$ff(x,y) \leftrightarrow |F(u,v)|^2 \tag{8}$$

where ff(x,y), the autocorrelation function (hereafter
referred to as the autocorrelation, for short) of the image,
is defined by

$$ff(x,y) = \iint f(\xi,\eta) \ f^*(\xi+x, \eta+y) \ d\xi \ d\eta \tag{9}$$

The autocorrelation box Bff, which is the square that just
encloses ff(x,y), is seen from (7) and (9) to be given by

$$Bff(x,y) = 1 \quad \text{for} \quad 0 < (|x|,|y|) < 1/N$$
$$= 0 \quad \text{for} \quad 1/N < (|x|,|y|) < \infty \tag{10}$$

It follows from (8) and the sampling theorem that $|F(u,v)|^2$
is completely determined by its samples spaced by N/2 in both
the u and v directions.

We assume from now on that the given samples of $|F(u,v)|$
are spaced no further apart than N/2, implying the complete
visibility magnitude can be reconstructed. With regard to
this, it is important to recognise that, in the real world,
there is often no precise prior estimate of the size of Bf
available to an experimenter. The practical connotations of
this are discussed in §12.

6. THE PERIODIC IMAGE AND THE CRYSTALLOGRAPHIC PHASE PROBLEM

We introduce the expanded image box, defined by

$$EBf(x,y) = 1 \quad \text{for } 0 < (|x|,|y|) < 1$$
$$= 0 \quad \text{for } 1 < (|x|,|y|) < \infty \qquad (11)$$

Note that the sides of EBf are N times those of Bf. the periodic image $p(x,y|N)$ is now defined by repeating the contents of EBf throughout all·of image space, in contiguous boxes each the size of EBf. We define the contents of EBf by, first, placing Bf in the middle of EBf, then filling Bf with $f(x,y)$ and, finally, setting the image amplitude to zero in the part of EBf not occupied by Bf. This somewhat eccentric construction facilitates subsequent analysis.

On defining $p(x,y|N) \leftrightarrow P(u,v|N)$, it follows that $P(u,v|N)$ is a two-dimensional array of two-dimensional delta functions at the points (m,n) in Fourier space, where m and n are arbitrary integers. Because of the way $p(x,y|N)$ is defined in the preceding paragraph, the complex amplitudes of the delta functions equal the $F(m,n)$. Since $P(u,v|N)$ is effectively zero except at the above-mentioned array of points, the $F(m,n)$ are the only values of $F(u,v)$ manifested by $P(u,v|N)$.

The atoms of which crystals are composed almost always "fill" the unit cell, in the sense that there is rarely much in the way of "empty space" in a crystal structure. This is equivalent to stating there is little point in considering values of N other than unity when invoking $p(x,y|N)$ to represent a (two-dimensional) crystal structure. The $F(m,n)$, which are what crystallographers would call the structure factors (cf. Ramachandran and Srinivasan 1970), are then spaced as far apart as is allowed by the sampling theorem - i.e. given the $F(m,n)$, we can reconstruct $f(x,y)$, but only just since, if the spacing of these samples of $F(u,v)$ was any wider, it would only be possible to recover $f(x,y)$ if we could devise other means for rematerialising the lost information. Crystallographers are not presented with the $F(m,n)$, however, but only with the $|F(m,n)|$. It is therefore impossible to reconstruct $ff(x,y)$ directly from the data, because to do so requires that we are presented with the $|F(m/2,n/2)|$, for all integers m and n spanning the region of Fourier space within which $|F(u,v)|$ has significant value, as is explained in §5 above.

Our approach to solving phase problems relies on being able to reconstruct (an estimate of) ff(x,y) immediately from the given data. Consequently, the crystallographic phase problem cannot be handled by the techniques described in this paper. We nevertheless feel that our methods may be of interest to crystallographers, for reasons outlined in §14.

7. SAMPLING GRIDS

For later convenience we need to introduce several rectangular grids (of points) in the square region R of Fourier space containing all the parts of $|F(u,v)|$ which exceed some prescribed threshold, the latter being set by the prevailing noise level. These grids identify sample points at which the magnitudes or complex amplitudes of samples of visibilities are given or specified. The names of the grids and the points which comprise them are listed below:

Primitive $(mN/2, nN/2)$
Actual (m,n)
In-line in-between $(m + 1/2,n)$ & $(m,n + 1/2)$
Diagonal in-between $(m + 1/2,n + 1/2)$

The integers m and n take on sufficient values in each case that the grids span R.

We can now paraphrase the penultimate paragraph of §6 more succinctly. In crystallography, the actual sampling grid coincides with those points of the primitive sampling grid for which m and n are both even. Only the actual samples of $|F(u,v)|$ can be observed, because $P(u,v|N)$ is effectively zero throughout Fourier space except on the actual sampling grid. In particular, the in-between crystallographic samples are "invisible". This emphasises that the "in-between-ness" is with respect to the actual samples.

For the in-line in-between sampling grid, we distinguish between sample points $(m + 1/2,n)$ in rows and $(m,n + 1/2)$ in columns.

8.   COMPATIBLE IMAGES AND MOST COMPACT IMAGES

Given the size of Bf, the size of Bff is deduced immediately, as indicated in §4 above - i.e. the sides of Bff are necessarily twice those of Bf.

We give the name compatible image to any image, say c(x,y), for which

$$cc(x,y) = ff(x,y) \tag{12}$$

It follows immediately from (8) that c(x,y) can be written as

$$c(x,y) \leftrightarrow |F(u,v)| \exp(i \; \Psi(u,v)) \tag{13}$$

where $\Psi(u,v)$ is any real "phase function" of u and v. We call the image-form (as defined in §3 above) of c(x,y) a compatible image-form.

If the size of Bff is given, it is obvious from inspection of the integral on RHS (9) that the sides of Bf cannot be shorter than one half of the sides of Bff. We say that any compatible image, whose image box is of this minimum size, is a most compact image, and its image-form is a most compact image-form.

It is clear from (13) that there must exist infinitely many image-forms whose image boxes exceed the minimum size. [It is not unlikely that all of these larger image boxes may strictly be of infinite size, but that is meaningless in any practical context, for which the image magnitude must fall below the prevailing noise level outside some effectively finite region of image space.] The phase function $\Psi(u,v)$ introduced in (13) forces c(x,y) to be complex in general. This implies that, when $c(\xi,\eta)$ is substituted for $f(\xi,\eta)$ in the integrand of RHS (9), the oscillations of $c(\xi,\eta)$ cause perfect cancellation of the integral for values of x and y corresponding to points lying outside Bff. The same thing can happen when $c(\xi,\eta)$ is real, provided it takes on negative as well as positive values. It is extremely difficult to imagine how this comes about, but it must do so of course because (8) is a mathematical theorem!

The only possible positive compatible images are most compact ones. While this result is transparent (Bates 1984; Bates and Fright 1984), it has such important consequences that it is worthy of being canonised as a theorem. It is

obvious because, when $f(\xi,\eta)$ is positive, none of the aforesaid cancellation can occur within the integral on RHS (9).

Since positive images are of interest in so many important technico-scientific applications (cf. Bates 1982b;
Bates, Fright and Norton 1984), it is very significant we can
be certain, for such images, that $Bf(x,y)$ must satisfy (7)
when the given $Bff(x,y)$ is defined by (10).

When the image is not necessarily positive, we can be
sure that the Fourier phase problem does not possess a unique
solution unless the two words "most compact" are inserted
before image-form in the first paragraph of §3. The crucial
question for all images is, then, how many most compact
image-forms can be compatible with a given visibility intensity?

9. PHASE FROM MAGNITUDE

We define the quasi-periodic image $e(x,y)$ by

$$e(x,y) = p(x,y|N) \, w(x) \, w(y) \qquad (14)$$

where $p(x,y|N)$ is introduced in §6, and where

$$\begin{aligned} w(x) &= (1 + \cos(\pi x))/2 \quad && \text{for } |x| < 1 \\ &= 0 \quad && \text{for } |x| > 1 \end{aligned} \qquad (15)$$

It follows from the convolution theorem (cf. Bracewell 1978)
that the Fourier transform of (14) is

$$E(u,v) = P(u,v|N) * S(u,v) \qquad (16)$$

where $P(u,v|N)$ is defined in §6, the symbol * denotes convo-
lution and

$$S(u,v) = S(u) \, S(v) \qquad (17)$$

with

$$S(u) = \text{sinc}(2u) + (\text{sinc}(2u + 1) + \text{sinc}(2u - 1))/2 \qquad (18)$$

where $\text{sinc}(t) = (\sin(\pi t))/\pi t$.

It follows from (17) and the definitions introduced in
§6 that (16) can be re-written as

$$E(u,v) = \sum_{m,n=-\infty}^{\infty} F_{m,n} \, S(u-m) \, S(v-n) \tag{19}$$

where $F_{m,n}$ is short-hand for the m,n^{th} actual sample of $F(u,v)$, i.e.

$$F_{m,n} = F(m,n) \tag{20}$$

The form of (18) ensures that

$$E_{m+1/2,n} = E(M + 1/2,n) = (F_{m+1,n} + F_{m,n})/2 \tag{21}$$

and

$$E_{m,n+1/2} = E(m,n + 1/2) = (F_{m,n+1} + F_{m,n})/2 \tag{22}$$

where the $E_{m+1/2,n}$ and $E_{m,n+1/2}$ are (short-hand for) the in-line in-between, in rows and columns respectively, samples of $E(u,v)$.

We now introduce notation for the magnitudes and phases of the actual samples of $F(u,v)$ – i.e. $A_{m,n}$ and $\Theta_{m,n}$ respectively –, and for the in-line in-between sample magnitudes of $E(u,v)$ – i.e. $B_{r,m,n}$ and $B_{c,m,n}$ with the letters r and c referring to rows and columns. So:

$$F_{m,n} = A_{m,n} \, \exp(i\Theta_{m,n}) \tag{23}$$

$$|F_{m+1/2,n}| = B_{r,m,n} \quad \text{and} \quad |F_{m,n+1/2}| = B_{c,m,n} \tag{24}$$

It follows immediately from (21) and (22) that

$$\cos(\Theta_{m+1,n} - \Theta_{m,n}) =$$
$$(4(B_{r,m,n})^2 - (A_{m+1,n})^2 - (A_{m,n})^2)/2 \, A_{m+1,n} \, A_{m,n} \tag{25}$$

and

$$\cos(\Theta_{m,n+1}, -\Theta_{m,n}) =$$
$$(4(B_{c,m,n})^2 - (A_{m,n+1})^2 - (A_{m,n})^2)/2 \, A_{m,n+1} \, A_{m,n} \tag{26}$$

We now show how to find the $\Theta_{m,n}$ given the $A_{m,n}$, the $B_{r,m,n}$ and the $B_{c,m,n}$. The connection between this phase problem and the Fourier phase problem (see §3) is established in §10.

Because we are concerned with the recovery of image-forms rather than images themselves (refer back to §3), we can arbitrarily prescribe the value of any one of the actual sample phases. We choose

$$\Theta_{0,0} = 0 \tag{27}$$

which, it is worth noting, is necessary when the image is

known to be positive. Substituting (27) into (25) for m = n
= 0 gives $\Theta1,0$ as the positive or negative arc-cosine of an
expression involving only the given quantities $Br,0,0$, $A1,0$
and $A0,0$. Because $f*(-x,-y)$ is defined (see §3) to have the
same image-form as $f(x,y)$, we can arbitrarily choose $\Theta1,0$ to
be positive:

$$\Theta1,0 = \mp arc \ cos((4(Br,0,0)^2 - (A1,0)^2 - (A0,0)^2)/2 \ A1,0 \ A0,0) \quad (28)$$

It is important to realise that (27) and (28) exhaust all the
flexibility inherent in the concept of the image-form. From
now on we must take explicit account of the two-fold ambiguity concerning the values of $(\Theta m+1,n - \Theta m,n)$ and
$(\Theta m,n+1 - \Theta m,n)$ implicit in (25) and (26) respectively.

Substituting (27) into (26) for m = n = 0 gives

$$\Theta0,1 = \pm\phi_1 \quad (29)$$

where ϕ_1 is the positive arc-cosine of an expression involving only given quantities - i.e. ϕ_1 is known. Substituting
(28) into (26) for m = (n+1) = 1 gives two possible values
for $\Theta1,1$ which, when substituted into (25) for m = (n-1) = 1,
gives

$$\Theta0,1 = \Theta1,1 \pm \phi_2 \quad (30)$$

where ϕ_2 has two known positive values, say $\phi_2{'}$ and ϕ_2 .
Unless either $|\phi_2{'} + \Theta1,0| = |\phi_2{`} - \Theta1,0|$, which is a special
case that almost never can occur, or $\Theta1,0 = 0$ or π, only one
of the two values specified by (29) and one of the four
values specified by (30) coincide. These coincident values
define $\Theta0,1$ unambiguously. When $\Theta1,0 = 0$ or π, the formula
(28) no longer represents a distinction between $f(x,y)$ and
$f*(-x,-y)$, implying that the positive, say, coincidence
between (29) and (30) can be chosen as the value of $\Theta0,1$ (it
is obvious that this argument can be extended until we come
across a $\Theta m,n$ which is neither 0 nor π).

The previous paragraph shows that, given two adjacent
actual sample phases, we can almost always unambiguously
reconstruct two further adjacent actual sample phases. The
latter two phases can serve as an initial pair for a repeat
of the procedure. Consequently, all the actual sample phases
can be recovered recursively in pairs.

This procedure, which was originally devised on a purely intuitive ad hoc basis (Bates (1982a), is reminiscent of Jennison's phase-closure technique, now routinely invoked by radio astronomers, and of the extension of speckle inter- ferometry due to Knox and Thompson (cf. §8.4 and §12.1, Bates 1982b).

10.   UNIQUENESS OF (THE MOST COMPACT) IMAGE-FORM

Given the primitive samples (see §7) of $|F(u,v)|$, we can immediately reconstruct $ff(x,y)$ and determine Bff (refer back to §5). By Fourier transformation of $ff(x,y)$, actual and in-line in-between samples of $|F(u,v)|$ can then be computed for any value of N.

Reference to the definition of $p(x,y|N)$ in §6 and to (14) confirms that

$$\lim_{N \to \infty} e(x,y) = f(x,y) \tag{31}$$

It follows that the Br,m,n and Bc,m,n are in-line in-between sample magnitudes of $F(u,v)$, as well as of $E(u,v)$, in the limit as $N \to \infty$. The phase recovery procedure described in §9 thus applies to $|F(u,v)|$ in this limit, indicating that there is almost always a unique solution to the Fourier phase prob- lem posed in §3. Since, as shown in §8, there are in general infinitely many compatible image-forms, this "unique" image- form must be the most compact one. This argument obviously extends to any greater number of dimensions than the two con- sidered explicitly here.

It therefore seems that there is almost always a unique most compact image-form. This is particularly significant when the image is known a priori to be positive, because (as shown in §8) any compatible image-form must then be most com- pact, implying that positive image-forms are almost always unique.

11.   FIENUP'S ALGORITHMS

We are now ready to discuss algorithms which have proven

capabilities for reconstructing image-forms from given visibility intensities. We only consider positive images from here on, because they occur in so many important scientific and technical applications, and most of the practical experience so far gained with phase recovery algorithms relates to positive image-forms.

Suppose that an estimate $\Psi(u,v)$ of phase$\{F(u,v)\}$ is available, where $\Psi(-u,-v) = -\Psi(u,v)$ necessarily because $f(x,y)$ is real. It can be combined with the given $|F(u,v)|$ to provide an estimate $\tilde{F}(u,v)$ for the visibility itself:

$$\tilde{F}(u,v) = |F(u,v)| \exp(i\ \Psi(u,v)) \tag{32}$$

We now define

$$\tilde{f}(x,y) \leftrightarrow \tilde{F}(u,v) \tag{33}$$

and denote by $f^+(x,y)$ the positive part of $\tilde{f}(x,y)$ inside the box $B = B(x-\xi, y-\eta)$, where B is unity inside a rectangle Ω in image space and zero outside it; Ω is centred on the point (ξ,η) and its sides are parallel to the x and y axes. There are several useful ways of choosing ξ and η which are described later. We write

$$\hat{f}(x,y) = \alpha\ f^+(x,y) - \beta\ f^-(x,y) \tag{34}$$

where α and β are positive constants, and $f^-(x,y)$ is the sum of two terms: the first is the negative part of $\tilde{f}(x,y)$ within Ω; the second is all of $\tilde{f}(x,y)$ outside Ω. On defining $\hat{F}(u,v)$ through

$$\hat{f}(x,y) \leftrightarrow \hat{F}(u,v) \tag{35}$$

it is convenient to re-define $\tilde{F}(u,v)$ by

$$\tilde{F}(u,v) = |F(u,v)| \exp(i\ \text{phase}\{\hat{F}(u,v)\}) \tag{36}$$

It is also convenient to retain the definition (33) but to extend the definition of $\hat{f}(x,y)$ to

$$\hat{f}(x,y) = \alpha\ f^+(x,y) - \beta\ f^-(x,y) + \gamma\hat{f}^`(x,y) + \delta\ \tilde{f}(x,y) \tag{37}$$

where the grave accent ` identifies a quantity from the immediately preceding iteration, γ is a positive constant and δ is a real constant.

An iterative loop is set up by applying (36), (37) and (35) in turn. The iterations are continued until either or

both of the error estimates εI or εF, defined below, become
less than pre-assigned values, chosen on the basis of the
estimated "noise" in the given (samples of) $|F(u,v)|$. The
labels I and F refer to image space and Fourier space respectively:

$$\varepsilon I = \iint\limits_{\Omega} |\hat{f}(x,y) - \tilde{f}(x,y)|^{\nu} \, dx \, dy \tag{38}$$

where ν is a positive number (usually taken to be 1 or 2),
and

$$\varepsilon F = \iint\limits_{R} ||F(u,v)| - |\tilde{F}(u,v)||^{\nu} \, du \, dv \tag{39}$$

where the region R of Fourier space is defined in §7. Note
that the integrand of (39) is identically zero at the actual
sample points, so that εF is a measure of how well the in-
between sample magnitudes are being reconstructed.

Fienup (1982) starts out with a pseudo-random positive
image confined to Bf. This is virtually equivalent to
employing a pseudo-random $\Psi(u,v)$. His simplest algorithm,
called error-correction or error-reduction, is characterised
by $\alpha=1$ and $\beta=\gamma=\delta=0$, with Ω being the same size as Bf and ξ
and η maximising the integral

$$\iint\limits_{\Omega} \tilde{f}(x,y) \, B(x-\xi,y-\eta) \, dx \, dy$$

which operation we refer to as "boxing the image". Fienup
begins his phase restoration by invoking the error-correction
algorithm, but after a number of iterations (based on inspection of the rate of decrease of εI) he usually changes over
to his hybrid-input-output algorithm, which is characterised
by $\alpha=\delta=0$, β equalling some appropriate number less than unity
(0.5 is typical), $\gamma = 1$, the values of ξ and η being fixed
where the error-correction algorithm sets them during its
final iteration, and the linear dimensions of Ω being about
10% greater than those of Bf. Fienup ceases iterating when
εI either falls below a pre-assigned value or merely exhibits
a tendency to oscillate. [While no mention is made in this
section of the fourth term on RHS (37), we suggest a use for
it in §14.]

12.   PRACTICAL PHASE RECOVERY

While the maximum entropy algorithm can be successful on

occasion (cf. Nityananda and Narayan 1982), the only phase
recovery procedures that can be said to have anything
approaching general, proven capabilities are all based on
Fienup´s algorithms.

Even though we of course rely on Fienup´s algorithms, we
are convinced that they need to be heavily supplemented when
one is attempting to recover image-forms from measured visibility intensities. We now list the steps which in our
experience (Bates and Fright 1983, 1984; Bates, Fright and
Norton 1984) maximise one´s chances of reconstructing a
recognisable version of an image-form from given primitive
samples (refer to §7) of its visibility intensity.

(i) Defogging

Our educated intuition compels us to assert that
Fienup´s algorithms are only efficient when $|F(u,v)|$ exhibits
appreciable interference. The visibility magnitudes
corresponding to many image-forms possess large central lobes
and comparatively small outer wiggles. The latter of course
carry the information about any interesting fine detail in
the images. An effective way of increasing the apparent
depth of the interference manifested by the wiggles is to
multiply $|F(u,v)|$ by a positive function, $M(u,v)$ say, which
is close to unity throughout most of R but dips smoothly
within the central lobe of $|F(u,v)|$ reducing its amplitude by
a factor, here denoted by the positive number k. Since the
central lobe corresponds to the background of the image,
which has the appearance of a "fog" when it is intense compared to the fine detail, its reduction is appropriately
called "defogging". When we refer from now on, at least
until step (vii), to $|F(u,v)|$, we actually mean ($|F(u,v)|$
$M(u,v)$).

(ii) Prefiltering

Because of the sampling and autocorrelation theorems
(refer to §5), when the primitive samples $|F(mN/2,nN/2)|^2$ of
the visibility intensity are input to an FFT (fast Fourier
transform algorithm, cf. Brigham 1974), the output is a set
of samples of ff(x,y). Inspection of the latter gives Bff,
so that the dimensions of Bf are immediately found by appealing to (7) and (10). Various practicalities must be noted.

Data are always contaminated with noise of one sort or
another. The FFT of the aforesaid primitive samples is thus
unavoidably different from ff(x,y). We call it aff(x,y). It
is necessary to estimate from inspection of aff(x,y) where
the periphery of Bff appears to lie. Furthermore, after dis-
carding all of aff(x,y) outside Bff, what remains is in general palpably different from zero on the periphery, in con-
tradiction of one's expectations of an autocorrelation function. We deem it appropriate to multiply aff(x,y) by a window function which is unity throughout most of Bff but falls
smoothly to zero at its periphery (we find that a product of
cosine bells, one a function of x and the other of y, is
satisfactory). The primitive samples of the Fourier
transform of the windowed aff(x,y) now serve as given data
for the remaining steps in our phase recovery procedure.
This prefiltering has two distinct advantages. The first is
that the effects of noise are ameliorated because the original data are smoothed, something which is particularly
appropriate when the data are derived from measured quantities. The second advantage, which is probably even more
important, is that the prefiltered primitive samples of the
visibility intensity are spaced as closely as demanded by the
linear dimensions of the Bf inferred from aff(x,y). So, we
are certain of the size of the image-form which is to be
reconstructed from the prefiltered $|F(u,v)|$. It is pointless
fretting that this is likely to differ from the image-form
implicit in the original data, because the latter is lost
irretrievably.

(iii)   Fringe magnification

While we have found defogging to be effective, even its
restorative powers need supplementing sometimes. Although it
emphasises the outer wiggles at the expense of the central
lobe of $|F(u,v)|$, it does nothing about minor fringes super-
imposed upon the wiggles. Faint detail in the image,
characterised by these minor fringes, is more readily
recovered (in our experience) if the fringes are enhanced.
We do this by forcing the function M(u,v), introduced in step
(ii), to exhibit a peak wherever a minor fringe manifests
itself. It is important for M(u,v) to remain smooth. We
arrange for all peaks in ($|F(u,v)|M(u,v)$) to fall off with
distance $\rho = (u^2 + v^2)^{\frac{1}{2}}$ from the origin of Fourier space as
$\exp(-\rho^2/2 \tau^2)$, where $\rho = 2\tau$ corresponds to where the u and v
axes intersect the periphery of R (refer to §7). When we
refer to $|F(u,v)|$ in step (iv) below we imply

$(|F(u,v)|M(u,v))$, with $M(u,v)$ defined as in this step.

(iv) Crude phase estimation

When operating on measured data, we often find that a
pseudo-random starting phase (refer to the final paragraph of
§11) is inadequate. The algorithm outlined in §9 can be
adapted to generate a much more satisfactory version of
$\Psi(u,v)$ from the visibility intensity. The quantities Br,m,n
and Bc,m,n introduced in §9 are taken to be the in-line in-
between samples of $|F(u,v)|$, rather than of $|E(u,v)|$ as
defined by (24). Because the prefiltered primitive samples
of $|F(u,v)|$ are spaced close enough (by definition) to com-
pletely characterise the prefiltered autocorrelation, we can
immediately compute all the samples of $|F(u,v)|^2$ defined in
§7. It is instructive to outline how this is done in a com-
putationally efficient manner. The primitive samples of
$|F(u,v)|^2$ are input to the FFT giving $aff(x,y)$ - which is now
considered to be $ff(x,y)$ as is intimated in step (ii) - which
is then packed with sufficient zeros that, when it is input
to the FFT, it produces the actual and in-between samples of
$|F(u,v)|^2$. We find that $N = 4$ is usually best for generating
the initial estimate $\Psi(u,v)$ of phase$\{F(u,v)\}$ by the procedure
described in the two paragraphs containing (27) through (30)
and in the paragraph following those two. [When phase information can be abstracted directly from measured data (even if
only very imperfectly) it may be preferable to use it to generate $\Psi(u,v)$ - Knox and Thompson's (1974) extension of
Laberyrie's (1970) speckle interferometry is a good example
of what might be done in an optical astronomical context.]

(v) Demagnification

While step (iii) helps to improve $\Psi(u,v)$, it often dis-
torts the visibility so much that the corresponding image-
form can no longer be positive. This prevents us invoking
Fienup's algorithms in their most convenient forms. Accord-
ingly, before proceeding to the next step, we set $|F(u,v)|$
equal to the version provided by step (ii).

(vi) Fienup cycling

We perform successive cycles, the jth consisting of J_j
iterations of the Fienup algorithms (see §11), using whatever
mix of error-correction and hybrid-input-output seems
appropriate (the choice is subjective, which is the most

unsatisfactory aspect of our current approach to phase restoration - a promising possibility for correcting this state
of affairs is discussed in §14). We assess the ongoing performance of the cycling by appealing to the error estimate
εF, defined by (38) - we feel it is more objective than εI
because its basis for comparison is the data, which are
represented by $|F(u,v)|$.

(vii) Refogging

Having obtained as good an estimate as possible of
phase$\{F(u,v)\}$ from step .(vi), we combine it with the original
given $|F(u,v)|$ - i.e. the version existing before any of the
steps are invoked. There is a practical point to be considered, however, whenever the given primitive samples appear
to be appreciably noisy. It may then be preferable to perform prefiltering, step (ii), without defogging, step (i), to
obtain smoothed primitive samples of $|F(u,v)|$. The latter
are combined with the phases generated by steps (i) through
(vi) above.

(viii) Final cycling

The version of F(u,v) obtained from step (vii) can sometimes be significantly improved by subjecting it to further
Fienup cycling.

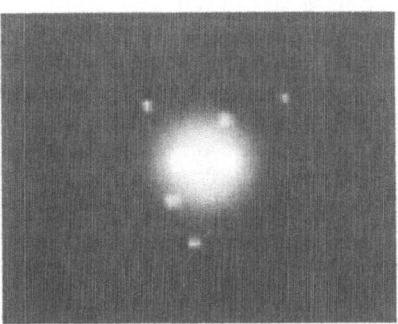

Figure 3

The true image f(x,y) for the computational examples
presented in §13.

 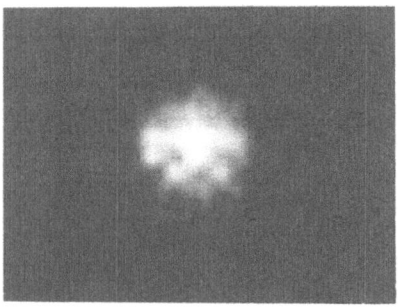

Figure 4

No defogging: $|F(u,v)|$ on the left and the
reconstructed image-form on the right.

13. COMPUTATIONAL EXAMPLES

The image shown in Figure 3 is to be taken as the true
image $f(x,y)$ for the examples presented in this section.
Their purpose is to illustrate the virtues of steps (i) and
(iii) of the procedure detailed in §12.

The image on the left in Figure 4 is the $|F(u,v)|$
corresponding to Figure 3. The image on the right in Figure
4 is the reconstructed image-form obtained using only steps
(iv) and (vi). During the latter step we performed 5 Fienup
cycles (each consisting of 20 iterations - 15 hybrid-input-
output, with $\beta = 0.5$, followed by 5 error-correction). While
some of the detail shown in Figure 3 is vaguely becoming
apparent in this reconstruction, it is still a long way from
being satisfactory.

The image on the left in Figure 5 is the previous visibility magnitude (shown in Fig. 4) after being subjected to
step (i) - i.e. defogging with $k = 0.05$. Note that the outer
wiggles (or sidelobes) of $|F(u,v)|$ are now much enhanced. On
invoking steps (iv) and (vi), we obtained the image-form

shown on the left in Figure 6 (we used 4 Fienup cycles, each
identical to those associated with Figure 4). A fair amount
of the detail shown in Figure 3 is apparent in this reconstruction. However, after invoking the refogging and final
cycling steps (one of the previously described Fienup cycles
for the latter step), (vii) and (viii) respectively, much of
the detail is obscured, as is indicated by the image on the
right in Figure 5.

The image on the left in Figure 7 is the previous visibility magnitude (shown in Fig. 5) after being subjected to
the fringe magnification step (iii). While this version of
$|F(u,v)|$ differs much less from that shown in Figure 5 than
the latter does from the version shown in Figure 4, the
degree of enhancement is sufficient to produce the image
shown on the right in figure 6, after invoking steps (iv),
(v) and (vi). This reconstruction is much better than the
one shown on the left. After invoking steps (vii) and (viii)
we obtain the image shown on the right in Figure 7. Its only
significant imperfection is a slight asymmetry of the fog (as
compared with Fig. 3).

The images shown in Figure 6 were both subjected to the
same amount of processing, except that the one on the right
was fringe-magnified and demagnified (both are computation-
ally efficient procedures, however). The images on the right
in both figures 5 and 7 were subjected to identical processing. We feel, therefore, that these examples confirm the
value of invoking fringe magnification as well as defogging.

There is another potential advantage of fringe magnification. It may help to remove the necessity for changing the
size of B during Fienup cycling, thereby making the whole
phase restoration procedure more objective (see the further
comments on this in §14).

14. CONCLUDING REMARKS

The only serious deficiency of the procedure outlined in
12 is that the strategies presently adopted during steps (vi)
and (viii) are quite subjective. It is absolutely vital to
develop entirely objective strategies if people are ever
going to be able to perform phase restoration routinely and

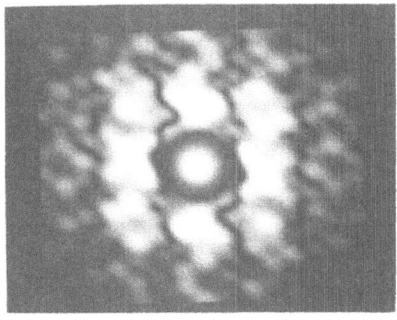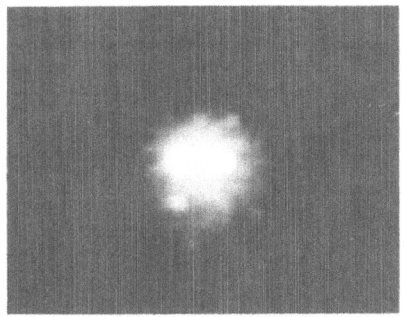

Figure 5

Defogging: $|F(u,v)|$ on the left and the refogged recon-
structed image-form on the right.

have confidence in the results. We feel a viable approach to
this problem is to modify Fienup cycling such that the parameters α, β, γ and δ appearing in (37) are adjusted on the
basis of what we call error-watching (Bates, Fright and Norton 1984).

The error estimate εF should be evaluated at each iteration and its rate of decrease (from iteration to iteration)
calculated. The parameters should be adaptively altered so
as to maximise the rate of decrease. It is not yet clear (to
us) how to do this efficiently, but we are presently studying
various possibilities.

Something else we are currently looking into is a way of
extending the crude phase estimation step (iv). The point
here is that the closer $\Psi(u,v)$ is to the true phase$\{F(u,v)\}$,
the more efficient should Fienup cycling be. It is also possible that an extended phase estimation procedure might be a
useful adjunct to the aforementioned algorithm-adjustment by
error-watching.

By taking advantage of the special conditions which per-
tain to their subject, X-ray crystallographers have developed
a marvellous kit of phase-restoration tools. The only structures that in any way tend to flummox them nowadays are

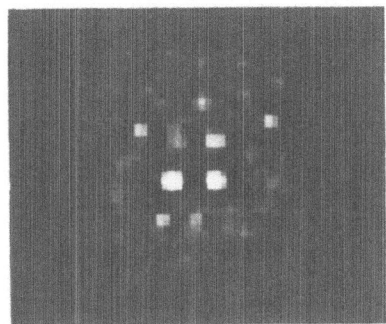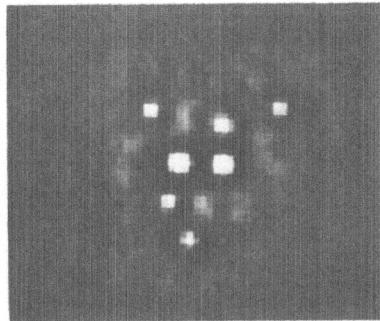

Figure 6

Defogged reconstructions of image-forms: without fringe
magnification on the left, but with it on the right.

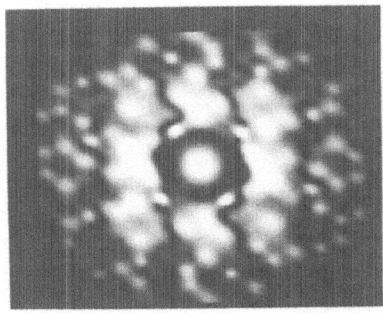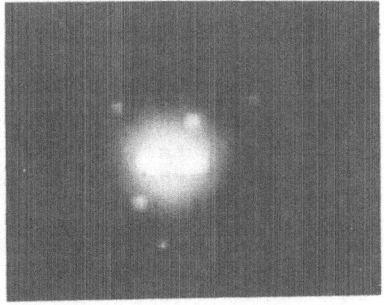

Figure 7

Defogging in combination with fringe magnification:
$|F(u,v)|$ on the left and the refogged reconstructed
image-form on the right.

macromolecules incorporating few heavy atoms, and for which
it is not feasible for one reason or another to synthesise
useful isomorphs. The structure factors of such molecules
can be "sharpened" (cf. Ramachandran and Srinivasan 1970),
which suggests that it just might prove possible to make useful estimates of the in-between structure factors. If we
could do this adequately, the structure factor phases could
be recovered straightforwardly using the procedure detailed
in §12.

REFERENCES

Bates, R.H.T., 1982a, Fourier phase problems are uniquely
        solvable in more than one dimension. I: underlying
        theory, Optik, 61: 247-262.

Bates, R.H.T., 1982b, Astronomical speckle imaging, Physics
        Reports, 90: 203-297. Bates, R.H.T., Uniqueness of
        solutions to two-dimensional Fourier phase problems
        for localized and positive images, Computer Graphics,
        Vision & Image Processing, 25(1984): 205-217.

Bates, R.H.T., and Fright, W.R., 1983, Composite two-
        dimensional phase-restoration procedure, J. Opt. Soc.
        America, 73: 358-365. Bates, R.H.T., and Fright,
        W.R., 1984, Reconstructing images from their Fourier
        "intensities", in: "Advances in Computer Vision and
        Image Processing Vol. 1" T.S. Huang, ed., J.A.I.
        Press, in press.

Bates, R.H.T., Fright, W.R., and Norton, W.A., 1984, Phase
        restoration is successful in the optical as well as
        the computational laboratory, in "Indirect Imaging",
        J.A. Roberts, ed. Cambridge Univ. Press, Cambridge,
        pp.119-124.

Bracewell, R.N., 1978, "Fourier Transform and its Applica-
        tions", McGraw-Hill (2nd edn), New York.

Brigham, E.D., 1974, "Fast Fourier Transform", Prentice-Hall,
        New Jersey.

Bruck, Y.M., Sodin, L.G., 1979, On the ambiguity of the image
        reconstruction problem, Optics Communications, 30:
        304-308.

Fiddy, M.A., Brames, B.J., and Dainty, J.C., 1983, Enforcing
    irreducibility for phase retrieval in two dimensions,
    Optics Letters, 8: 96-98.

Fienup, J.R., 1982, Phase retrieval algorithms: a comparison,
    Applied Optics, 21: 2758-2769.

Fienup, J.R., 1984, Experimental evidence of the uniqueness
    of phase retrieval from intensity data, in: "Indirect
    Imaging", J.A. Roberts, ed., Cambridge Univ. Press,
    Cambridge, pp.99-109.

Fright, W.R., Bates, R.H.T., 1982, Fourier phase problems are
    uniquely solvable in more than one dimension. III:
    Computational examples for two dimensions, Optik, 62:
    219-230.

Gull, S.F., Daniell, G.J., 1978, Image reconstruction from
    incomplete and noisy data, Nature, 272: 686-690.

Huiser, A.M.J., van Toorn, P., 1980, Ambiguity of the phase-
    reconstruction problem, Optics Letters, 5: 499-501.

Kiedron, P., 1981, On the 2-D solution ambiguity of the phase
    recovery problem, Optik, 59: 303-309.

Knox, K.T., and Thompson, B.J., 1974, Recovery of images from
    atmospherically-degraded short-exposure photographs,
    Astrophysical J., 193: L45-L48.

Labeyrie, A., 1970, Attainment of diffraction-limited resolu-
    tion in large telescopes by Fourier analysing speckle
    patterns in star images, Astronomy & Astrophysics, 6:
    85-87.

Napier, P.J., Bates, R.H.T., 1974, Inferring phase informa-
    tion from modulus information in two-dimensional aper-
    ture synthesis, Astronomy & Astrophysics Supplement
    Series, 15: 427-430.

Nityananda, R., and Narayan, R., 1982, Maximum entropy image
    reconstruction - a practical non-information-theoretic
    approach, J. Astrophysics & Astronomy, 3: 419-450.

Ramachandran, G., and Srinivasan, R., 1970, "Fourier Methods
    in Crystallography", Wiley, New York.

Taylor, L.S., 1981, The phase retrieval problem, IEEE Tran-
    sactions, AP-29: 386-391.

FOURIER UNIQUENESS CRITERIA AND SPECTRUM ESTIMATION THEOREMS

John J. Benedetto

Department of Mathematics
University of Maryland
College Park
Maryland   USA

The convergence or divergence of logarithmic integrals,

$$\int_{-\infty}^{\infty} \frac{|\log F(\gamma)|}{1 + \gamma^2} \, d\gamma, \tag{1}$$

is related to entropy and periodogram spectrum estimation methods, respectively.

Periodogram spectrum estimation is based on the existence of a unique power spectrum which must be approximated; the proof of existence depends on the divergence of integrals (1).   Entropy spectrum estimation is based on modelling eligible power spectra subject to maximizing the entropy rate of corresponding processes; the class of such power spectra depends on the convergence of integrals (1). Topics in this paper include the Rieszes´ and Levinson´s uniqueness theorems, the Szegö factorization theorem, the work of Paley and Wiener, the spectral analysis of non-quasi-analytic functions, and the Beurling–Malliavin theory. The results are all related to (1), and they are presented in the context of their common spectrum estimation background.

Our notation follows [2] in this volume.

1. PERIODOGRAM SPECTRUM ESTIMATION AND UNIQUENESS THEOREMS

In light of periodogram spectrum estimation algorithms we consider the following point of view. Let R_1 and R_2 be two autocorrelation extensions to the real line R of given autocorrelation data on a fixed time interval $[-T,T]$, and set $\hat{S}_j = R_j$, $R_{12} = R_1 - R_2$, and $\hat{S}_{12} = R_{12}$. If the support of S_{12}, denoted by supp S_{12}, is compact for each S_1 and S_2 then $R_{12} = 0$ on R since, in this case, not only does R_{12} vanish on the interval $[-T,T]$ but also R_{12} is analytic. Thus the power spectrum of the given auto-correlation data on $[-T,T]$ is <u>uniquely</u> defined, and the periodogram algorithms aim to approximate S in the best possible way.

In case we only know that supp $S_{12} \subseteq [-\Omega,\infty)$ for each S_1 and S_2 then the above argument defines a unique power spectrum if the abscissa of convergence σ_{12} of the Laplace transform of each S_{12} is always negative. If $\sigma_{12} = 0$ for some S_{12} then the existence of a <u>unique</u> power spectrum depends on the following theorem which is due to F. and M. Riesz: the Laplace transform of the bounded Radon measure S_{12} is (easily seen to be) a bounded analytic function in the half-plane Re $z > 0$, and so $S_{12} = 0$ if the Lebesgue measure of $\{t : \hat{S}_{12}(t) = 0\}$ is positive, e.g., E. Hille, Analytic function theory, Section 19.

We reiterate that the uniqueness property is prerequisite for periodogram methods. We formalize this in the following way.

<u>Theorem 1.1.</u> Given a function $x : [-T,T] \times P \to C$, where P is a probability space. If the condition,

for all y, there exists Ω_y such that supp $S_y \subseteq [-\Omega_y,\infty)$,

is satisfied, where y is an extension of x to $R \times P$ (that is, y is a stationary stochastic process and $y = x$ on $[-T,T] \times P$), then there is a uniquely determined positive measure S which is the power spectrum of every extension y of x. The measure S is not necessarily computable, e.g., <u>JMAA</u> 91 (1983) 444-509, Section III.1.

<u>Proof.</u>   a.   Let y_1 and y_2 be extensions of x where

$$\text{supp } S_{y_j} \subseteq [-\Omega_j, \infty), \quad \Omega_j \geqslant 0 .$$

Let $b = \chi_T$ so that $b*b > 0$ on $(-2T, 2T)$.   Clearly, for $j = 1, 2$,

$$\text{for all } \omega, \quad E\{S_b\}(\omega) = S_{y_j} |\tilde{B}|^2 (-\omega),$$

since $x = y_1 = y_2$ on $[-T, T] \times P$; and, consequently,

$$(R_{y_1} - R_{y_2})(b*b) = 0 \quad \text{on} \quad R,$$

where $\hat{S}_{y_j} = R_{y_j}$.   Because $b*b > 0$ on $(-2T, 2T)$ we conclude that

$$\text{for all } t \in (-2T, 2T), \quad R_{y_1}(t) = R_{y_2}(t). \tag{1.1}$$

b.   Let $\hat{S}_{12} = R_{12} = R_{y_1} - R_{y_2} = 0$ on $(-2T, 2T)$,
where supp $S_{12} \subseteq [-\Omega, \infty)$ and $\Omega = \max(\Omega_1^2, \Omega_2)$.   By the
Rieszes' result, we conclude that $S_{12} = 0$.

q.e.d.

<u>Remark 1.1.</u>   a.   If R is real then $\overline{S(-\gamma)} = S(\gamma)$.   In
particular, the support hypotheses of Theorem 1.1 are not
viable.   The point of Theorem 1.1, and the remarks preceding
it, is to establish a mathematical basis on which to build
the periodogram algorithms of spectrum estimation.   A completely rigorous foundation is not possible, but uniqueness
in the spirit of Theorem 1.1 is a fundamental mathematical
component in any attempt to justify the subsequent algorithms.

b.   An essential feature of any periodogram
method is to begin by smoothing the periodogram.   This
apparently bad idea seems necessary in order to exhibit the
unique power spectrum S which must then be estimated by the

given periodogram algorithm.

The "$H^1(R)$" version of the Rieszes' result is

Theorem 1.2 (Rieszes). If $R \varepsilon L^1(R)$, $|\{t: R(t) = 0\}| > 0$, and supp $S \subseteq [-\Omega,\infty)$, where $\hat{S} = R$, then $R = 0$ ($|X|$ is the Lebesgue measure of X).

For proofs, old and new, and extensions of the Rieszes' theorem we refer to [18], [15], and [3], respectively.

Analyticity has played a role in the previous discussion on uniqueness. Levinson proved results analogous to the Rieszes' without explicit mention of analytic functions. The following general form of Levinson's theorem is due to deBranges, cf., [20, Chapter V].

Theorem 1.3. Let $K \geqslant 1$ be a continuous function on R for which log K is uniformly continuous and

$$I(K) = \int_{-\infty}^{\infty} \frac{\log K(\gamma)}{1 + \gamma^2} \, d\gamma = \infty.$$ (1.2)

If $S \varepsilon M(R)$ satisfies the conditions

$$\int_{-\infty}^{\infty} Kd|S| < \infty$$ (1.3)

and

$$R = 0 \text{ on an interval,}$$ (1.4)

where $\hat{S} = R$, then $S = 0$.

Example 1.1. The Rieszes' theorem is a corollary of Levinson's theorem if $R \varepsilon L^1(R) \cap M(R)^\wedge$ and $R = 0$ on an interval. In fact, if we define

$$K(\omega) = \begin{cases} 1, & \omega < 0 \\ e^{-\omega}, & \omega > 0 \ , \end{cases}$$

and if the hypotheses of Theorem 1.2 hold, then $I(K) = \infty$ and (1.3) is valid since supp $S \subseteq [-\Omega,\infty)$. Thus, Levinson's theorem applies and $S = 0$.

The primordial case of Levinson's time-gap theorem is

Theorem 1.4. Suppose $S \in L^1(\mathbb{R})$ and there are constants $C, c > 0$ for which $|S(\gamma)| < C \exp(c\gamma)$ for all $\gamma < 0$. If $\hat{R} = S$ and R vanishes on an interval, then $S = 0$.

Proof Taking $z = t + iu$ we have

$$|R(z)| < |\int_{-\infty}^0| + |\int_0^\infty S(\gamma)e^{2\pi i(t+iu)\gamma} \, d\gamma|.$$

The first integral is bounded by

$$C \int_{-\infty}^0 e^{\gamma(c-2\pi u)} \, d\gamma,$$

which converges for all $u < c/(2\pi)$. The second integral clearly converges for all $u > 0$.

Thus $R(z)$ exists on the strip Im $z \in [0, c/2\pi)$. The same result holds for $R'(z)$ and so we can conclude that R is analytic in the strip Im $z \in (0, c/2\pi)$. Since $R = 0$ on an interval there are classical methods to ensure that $S = 0$, e.g. [3].

q.e.d.

Example 1.2. Theorem 1.4 is seen to be a corollary of Levinson's theorem by taking

$$K(\omega) = \begin{cases} 1, & \omega > 0 \\ e^{b\omega}, & \omega < 0 \end{cases},$$

for some $b \varepsilon (-c,0)$.

Remarks 1.2.   We can use the simple proof of Theorem 1.4 to obtain Levinson's theorem in the case (1.2) is replaced by the condition,

$$\text{there exists } r > 0 \text{ such that } \sup_{\gamma < 0} \frac{|\gamma| e^{|\gamma| r}}{K(\gamma)} < \infty. \quad (1.5)$$

In fact, we compute

$$|R(z)| < \int_{-\infty}^{0} \frac{e^{-2\pi u \gamma}}{K(\gamma)} K(\gamma) d|S|(\gamma) + \int_{0}^{\infty} e^{-2\pi u \gamma} d|S|(\gamma)$$

and we have convergence in a strip by (1.5), (1.3), and the hypothesis $S \varepsilon M(\mathbb{R})$.

Clearly (1.5) implies (1.2).   In fact, (1.5) is valid if and only if $\sup\{\log|\gamma| + |\gamma|r - \log K(\gamma) : \gamma < 0\} < C$, and this condition is characterized by the inequality,

$$\text{for all } \gamma < 0, \ \frac{\log|\gamma| + |\gamma|r}{1+\gamma^2} < \frac{C}{1+\gamma^2} + \frac{\log K(\gamma)}{1+\gamma^2},$$

which, in turn, implies (1.2).

Beurling [5] weakened condition (1.4) to the condition, $|\{t : R(t) = 0\}| > 0$, in several important cases of the uniqueness theory centering about Levinson's theorem.   He also proved versions of Levinson's theorem which are easier to implement than Theorem 1.3.   For example, he proved the following for tempered distributions.

Theorem 1.5.   Given $S \varepsilon M(\mathbb{R})$ for which (1.4) is satisfied.   If the complement of supp S contains a set

$\cup \, [k_n - r_n, \, k_n + r_n]$, where $0 < k_1 < k_2 < \cdots, \; k_n - r_n > k_{n-1}$

for $r_n > 0$, and

$$\Sigma (r_n / k_n)^2 = \infty,$$

then $S = 0$.

Example 1.3.   Taking supp $S \subseteq (-\infty, 0]$ and $R = 0$ on an interval we see that the Rieszes' theorem is a corollary of Theorem 1.5 by noting that $(0, \infty)$ contains $\cup [k_n - r_n, \, k_n + r_n]$, where $k_n = n + 1$ and $r_n = n^{1/2}$.

The latest word on these matters is contained in the important work of Benedicks [3].

2.   THE FACTORIZATION THEOREM

$T = R/Z$ is identified with the interval $[0, 1)$. $H^p(T)$, $p \in [1, \infty)$, is the set of Fourier series

$$F \sim \sum_0^\infty f_j \exp(2\pi i j \gamma),$$

analytic in the unit disc $|z| < 1$, for which $\sum_0^\infty |f_j|^p < \infty$.

Theorem 2.1   (Szegö factorization).   Let $G \in L^1(T) \backslash \{0\}$ and assume $G \geq 0$ on T. $G = G(0) |G_+|^2$ a.e., where $G_+ \in H^2(T)$ and $|G_+(0)| = 1$, if and only if $\log G \in L^1(T)$. Further, $G_+(z) \neq 0$ for all $|z| < 1$.

Szegö's original proof of Theorem 2.1 (Math. Ann. 84 (1921) 232–244) depended on his prediction theory "Szegö alternative" theorem (Math. Zeit. 6 (1920) 167–202).

Theorem 2.2   (Szegö alternative).   Let $W \varepsilon L^1(T)$ be
non-negative and define the __geometric mean__ of W   as

$$
g(W) = \begin{cases} \exp \int_T \log W(\gamma)d\gamma, & \log W \varepsilon L^1(T) \\ 0 \quad, & \log W \notin L^1(T) . \end{cases}
$$

Then

$$
\inf \int |1 - P(\gamma)|^2 W(\gamma)d\gamma = g(W),
$$

where the infimum is taken over all trigonometric polynomials
P on T with vanishing constant term.

Theorem 2.2 has been generalized in several directions,
e.g., [1, pp. 256 ff.; 15, pp.48-50].

The proof of Theorem 2.1 in [12, Chapter 1.14] is joint
work of F. Riesz and Szegö which actually appeared before
Szegö's original proof, cf. [22, pp. 115 ff.].

For the real line R we have [23, Theorem XII] –

Theorem 2.3 (Paley-Wiener).   Let $G \varepsilon L^2(R) \setminus \{0\}$ and assume
$G > 0$ on R. $G = |G_+|$ a.e., where $G_+ \varepsilon L^2(R)$ and
supp $\hat{G}_+ \subseteq [0,\infty)$, if and only if

$$
\int_{-\infty}^{\infty} \frac{|\log G(\gamma)|}{1 + \gamma^2} d\gamma < \infty. \tag{2.1}
$$

Further, $G_+(z)$ is analytic and non-zero in the open upper-
half plane Im z > 0.

Remark 2.1.   a. G_+ is defined in the following way which
shows explicitly that it is non-vanishing for Im z > 0.
First, consider the harmonic function,

$$
u(x + iy) = \frac{1}{\pi} \int_{-\infty}^{\infty} \frac{y \log G(\gamma)}{(x - \gamma)^2 + y^2} d\gamma, \quad y > 0,
$$

and let $v(x + iy)$ be its harmonic conjugate. Next, set $G_+(z) = \exp(u(z) + iv(z))$, $z = x + iy$ and $y > 0$.

b. Using Theorem 2.3 we obtain the analogue of Theorem 2.1 for $\mathbb{R}$ as follows. Let $G \in L^1(\mathbb{R})_{1/2} \setminus \{0\}$ be non-negative and assume (2.1). Setting $F = G^{1/2}$, we see that $F \in L^2(\mathbb{R}) \setminus \{0\}$ and (2.1), with G replaced by F, is valid. Consequently, we apply Theorem 2.3 to obtain $F_+ \in L^2(\mathbb{R})$ such that $|F_+| = F$ a.e. and supp $F_+ \subseteq [0,\infty)$. The desired factorization is $G = |F_+|^2$.

By considering $D_n \hat{\mu}$ where $\hat{d}_n = D_n$ and $d_n = n\chi_{1/(2n)}$, we can use Theorem 2.3 to verify easily that:

Theorem 2.4. Given $\mu \in M(\mathbb{R})$. Then supp $\mu \subseteq [T,\infty)$ if and only if

$$\int_{-\infty}^{\infty} \frac{|\log|\hat{\mu}(\gamma)||}{1 + \gamma^2} \, d\gamma < \infty. \qquad (2.2)$$

Thus, (2.2) completely describes how fast the Fourier transforms of measures supported on half-lines can vanish at $+\infty$. For compactly supported measures the situation is more complicated [6; 21; 17].

Theorem 2.5 (Beurling-Malliavin). Let $|K| > 1$ be a continuous function on $\mathbb{R}$ for which $\log|K|$ is uniformly continuous. The condition,

$$\int_{-\infty}^{\infty} \frac{\log|K(\gamma)|}{1 + \gamma^2} \, d\gamma < \infty, \qquad (2.3)$$

is necessary and sufficient that for all $\varepsilon > 0$ there exists $\mu \in M(\mathbb{R})$ such that supp $\mu \subseteq [-\varepsilon,\varepsilon]$ and $\hat{\mu}K \in L^\infty(\mathbb{R})$.

The necessity of (2.3) can be verified using Theorem 2.4. The sufficiency is very difficult and depends heavily on properties of entire functions of exponential type.

Example 2.1. Because we are interested in the integrability of $\log G$ on T and condition (2.1) on $\mathbb{R}$, we make the following trivial observations. a) If $G > 0$ on T and

$G \in L^1(T)$ then $\log G \in L^1(T)$. b) In Theorem 2.1, we see that $\log|G_+| \in L^1(T)$ when $\log G \in L^1(T)$. c) If $G > 0$ and $G \in L^1(T)$ then $\log G \in L^1(T)$ if and only if $\int_T \log G(\gamma)d\gamma > -\infty$. This last point follows since

$$\int_{G>1} |\log G| = \int_{G>1} \log G < \int_{G>1} G$$

and

$$\int_{G<1} |\log G| = - \int_{G<1} \log G$$

for $G > 0$; in fact, given non-negative $G \in L^1(T)$,

$$\log G \in L^1(T) \leftrightarrow \int_{G<1} \log G < \infty$$
$$\leftrightarrow \int_{G<1} \log G > -\infty \leftrightarrow \int \log G > -\infty.$$

Given $F \in H^1(T)$ for which $|\{\gamma : F(\gamma) = 0\}| = 0$. It is well known, e.g., [15, p. 52], that $\log F \in L^1(T)$ and

$$\int_T \log|F(e^{i\gamma})|d\gamma > \log|F(0)|; \tag{2.4}$$

in case of equality in (2.4) we say that F is an <u>outer function</u>. Given $G \in L^1(T)$ which is positive on T. If in Szegö's factorization $G = G(0)|G_+|^2$, which is unique [12, p. 26], $G_+ \in H^1(T)$ is an outer function, then we say that G is an <u>outer modular function</u>.

<u>Example 2.2.</u> If G is a positive continuous function on T then G is an outer modular function. To see this let $G = G(0)|G_+|^2$ be the Szegö factorization, and note that $\log|G_+|$ is continuous on T and harmonic in $|z| < 1$ since $G_+(z) \neq 0$ for $|z| < 1$. The result follows since, by the continuity of $\log|G_+|$ on the closed unit disc, we can write $\log|G_+|$ as a Poisson integral for all $|z| < 1$.

3. THE MAXIMUM ENTROPY THEOREM

In [2] of this volume we proved the maximum entropy theorem MEM. We now state a generalization, which is not extensive but which emphasizes the role of the convergent integrals, $\int \log G(\gamma)d\gamma < \infty$, discussed in Section 2.

Recall that a matrix $R = (r_{ij})$ is <u>hermitian</u> if $r_{ij} = \bar{r}_{ji}$. If $x = (x_0, \ldots, x_n)$ and $y = (y_0, \ldots, y_n)$ then we define

$$\langle Rx,y \rangle = \sum_{i,j} r_{ij} x_j \bar{y}_i.$$

An hermitian matrix R is <u>positive definite</u>, written $R \gg 0$, if $\langle Rx,x \rangle > 0$ for all x and $\langle Rx,x \rangle = 0$ only when $x_0 = x_1 = \ldots = x_n = 0$.

<u>Example 3.1.</u>   a. $R = (r_{ij})$ is hermitian if and only if $\langle Rx,y \rangle = \langle x,Ry \rangle$ for all x,y.

b.   If $R = (r_{ij})$ is hermitian, then each eigenvalue γ_j, $j = 0, \ldots, n$, of R is real.

c.   If $R = (r_{ij})$ is hermitian then $R \gg 0$ if and only if each eigenvalue γ_j, $j = 0, \ldots, n$, of R is positive.

d.   If $R = (r_{ij})$ is hermitian and R^{-1} exists then R^{-1} is hermitian.

<u>Example 3.2.</u>   Given $\{r_j : j = 0, \pm 1, \ldots, \pm n$ and $\bar{r}_j = r_{-j}\}$ and set $R = (r_{ij}) = (r_{i-j})$.
a. R is hermitian.

b.   If $R \gg 0$ then $r_0 > 0$.

c.   If $R \gg 0$ then $\det R > 0$.

d.   If $R \gg 0$ then $r_0 > |r_k|$ for $1 < k < n$.

Example 3.3.   a.   Szegö's factorization theorem (Theorem
2.1) can be considered an extension of the following classical result of Fejér-Riesz [12, Section 1.12; 24, pp. 231-
234]:   if

$$G = \Sigma^n_{-n} \, g_j \exp(2\pi i j\gamma) \geqslant 0$$

on   T   then there is a polynomial

$$G_+ = \Sigma^r_0 \, g^+_j \, \exp(2\pi i j\gamma)$$

for which $G = |G_+|^2$ on   T;

               b.   For any trigonometric series
$G \sim \Sigma \, g_j \exp(2\pi i j\gamma)$ we define $R(G) = \Sigma^n_{-n} \, g_j \bar{r}_j$ where
$\{r_j : j = 0, \pm 1, \ldots, \pm n$ and $\bar{r}_j = r_{-j}\}$ is given.   R can
be viewed as an operator on the space of trigonometric
series.   Clearly, if

$$G_+ \sim \Sigma^n_0 \, g^+_j \, e^{2\pi i j\gamma} \quad \text{then} \quad R(|G_+|^2) = \Sigma_{i,j} \, r_{i-j} g^+_j \bar{g}^+_i.$$

The operator   R   is _positive_ if for all

$$G \sim \Sigma^n_{-n} \, g_j e^{2\pi i j\gamma}$$

we have $R(G) \geqslant 0$ and $R(G) = 0$ only when $G = 0$.

               c.   Given $\{r_j : j = 0, \pm 1, \ldots, \pm n$ and $\bar{r}_j = r_{-j}\}$.   Define the matrix $R = (r_{i-j})$ and the operator   R.
Then the matrix   R   is positive definite if and only if the
operator   R   is positive.   The verification that   R   is
positive requires the Fejér-Riesz theorem.

     Let   A(T)   denote the space of absolutely convergent
Fourier series on   T.

Theorem 3.1 (MEM).   Given $\{r_j : j = 0, \pm 1, \ldots, \pm n$ and
$\bar{r}_j = r_{-j}\}$ and assume $R = (r_{i-j}) \gg 0$.   There is a unique
absolutely convergent Fourier series $S \sim \Sigma \, s_j \exp(2\pi i j\gamma)$,

$\hat{S}(j) = s_j$, with the following properties:

a. for all $|j| < n$, $s_j = r_j$;

b. $S > 0$ on T and $S^{-1} \in A(T)$;

c. $S = |S_+|^2$ on T where $S_+ \sim \sum_0^\infty s_j^+ e^{2\pi i j \gamma} \in A(T)$;

d. for all $|j| > n$, $(S^{-1})^\wedge(j) = 0$ and

$$S(e^{2\pi i \gamma}) = 1/(\sum_{k=-n}^{n} (\frac{1}{c_{00}} \sum_{j=0}^{n} \bar{c}_{0j} c_{0,j-k}) e^{2\pi i k \gamma}),$$

where $R^{-1} = (c_{ij})$;

e. for all $G \in L^1(T)$, for which $G > 0$ on T , G is
    an outer modular function, and $\hat{G}(j) = r_j$ for all
    $|j| > n$, we have

$$\int_T \log G \, d\gamma < \int_T \log S \, d\gamma;$$

and equality is obtained if and only if $G = S$.

4.   KOLMOGOROFF'S AND BEURLING'S THEOREMS FOR WEIGHTED L^1

As we noted in [2] of this volume, we have

$$E\{x(t_1) \, \overline{x(t_2)}\} = \int_{-\infty}^{\infty} e^{-2\pi i \gamma(t_1 - t_2)} \, dS(\gamma)$$

for a stationary stochastic process x with power spectrum
S.   Thus the mapping,

$$e^{2\pi i \gamma t} \to x(t, \alpha),$$

establishes an isomorphism between $L_S^2(R)$ and the span of
$\{x(t, \cdot) : t \in R\}$ completed with the norm

$$(\int |x(t,\alpha)|^2 \, dp(\alpha))^{1/2};$$

$L_S^2(\mathbb{R})$ is the Hilbert space of measurable functions F for which $\int |F(\gamma)|^2 \, dS(\gamma) < \infty$. As we pointed out, this isomorphism is fundamental in prediction theory in the case that the statistics of x are specified by the autocorrelation, cf., the article by Muhly in [27, pp. 340 ff.], the article by Masani in [27, pp. 276-306], the article by Kailath in [16], [13], and [10]. The basic point of view is that closure theorems in $L_S^p(\mathbb{R})$ give prediction theoretic information about the stochastic process. A fundamental closure theorem due to Kolmogoroff (on $\mathbb{T}$) and Krein, originally depending on the Szegö alternative Theorem 2.2, is the following.

Theorem 4.1.   Let $S \in M(\mathbb{R})$ be a positive measure with absolutely continuous part $F \in L^1(\mathbb{R})$, and let

$$L_S^p(\mathbb{R}) = \{G : \|G\|_{p,S} = (\int_{-\infty}^{\infty} |G(\gamma)|^p \, dS(\gamma))^{1/p} < \infty\},$$

$p \in [1,\infty)$ fixed.   Then the set $\{\exp(2\pi i \gamma t) : t < 0\}$ spans $L_S^p(\mathbb{R})$ taken with norm $\| \ \|_{p,S}$ if and only if

$$\int_{-\infty}^{\infty} \frac{|\log F(\gamma)|}{1+\gamma^2} \, d\gamma = \infty, \tag{4.1}$$

e.g., [1, pp. 261 ff; 15, pp. 48-50] for proofs on $\mathbb{T}$.

Note that the closure condition depends only on F.

For the case that $F_1 = K > 1$ (in particular, $K \notin L^1(\mathbb{R})$ whereas $F \in L^1(\mathbb{R})$ in Theorem 4.1) Beurling [4] considered

$$L_K^1(\mathbb{R}) = \{G \in L^1(\mathbb{R}) : \|G\|_{1,K} = \int_{-\infty}^{\infty} |G(\gamma)| K(\gamma) d\gamma < \infty\}$$

where K is even, $1 = K(0) < K(\gamma)$ and $K(\gamma + \omega) < K(\gamma)K(\omega)$. As such, $L_K^1(\mathbb{R})$ is a convolution algebra and Beurling posed the spectral analysis question: is every proper closed ideal $I \subseteq L_K^1(\mathbb{R})$ contained in a regular (i.e., $L_K^1(\mathbb{R})/I$ has a unit) maximal ideal?   For an equivalent analytic means of posing this question consider the following property of $I = L_K^1(\mathbb{R})$:

for all $t \in \mathbb{R}$ there exists $F \in I$ (4.2)

such that $f(t) \neq 0$, $\hat{F} = f$.

Then the spectral analysis question is equivalent to finding conditions on K so that whenever a closed ideal I satisfies (4.2) we can conclude that $I = L_K^1(\mathbb{R})$. Beurling's basic closure theroem in this direction is

<u>Theorem 4.2</u>   (Beurling).   Given $L_K^1(\mathbb{R})$ when K satisfies

$$\int_{-\infty}^{\infty} \frac{\log K(\gamma)}{1 + \gamma^2} d\gamma < \infty.$$
    (4.3)

Then the spectral analysis question has an affirmative answer.

<u>Example 4.1</u>   Let $K = 1$ on $\mathbb{R}$.   Then (4.3) is satisfied and Theorem 4.2 is Wiener's Tauberian theorem: $F \in L^1(\mathbb{R})$ has a non-vanishing Fourier transform if and only if the closed principal ideal $I_F \subseteq L^1(\mathbb{R})$ generated by F is all of $L^1(\mathbb{R})$.

<u>Remark 4.1.</u>   a.   Beurling viewed Theorem 4.2 in terms of non-quasi-analystic classes for the following reason.   Given $k_0 = 1, k_1, k_2, \ldots > 0$ for which $k_n^2 < k_{n-1} k_{n+1}$, e.g., $k_n = n!$.   Let $C\{k_n\}$ be the set of bounded infinitely differentiable functions F with the property that there are constants $c(F)$, $k(F)$ such that

for all $n \geq 1$, $\|F^{(n)}\|_\infty < c(F) k(F)^n k_n$.

$C\{k_n\}$ is a quasi-analytic class if, for all $F \in C\{k_n\}$, the condition $F^{(n)}(0) = 0$ for all n implies $F = 0$.   (If F is analytic and bounded by $c(F)$ in an open disc of center 0 and radius $R = 1/k$, then $|F^{(n)}(0)| < c(F) k^n n!$ for each $n \geq 1$.)   Setting

$$K(\gamma) = \sup_{n \geqslant 0} |\gamma|^n / k_n \geqslant 1 \quad \text{for} \quad \gamma \neq 0,$$

the Denjoy-Carleman theorem asserts that $C\{k_n\}$ is a quasi-analytic class if and only if

$$\int_{-\infty}^{\infty} \frac{\log K(\gamma)}{1 + \gamma^2} d\gamma = \infty, \qquad (4.4)$$

e.g., Ostrowski in Acta Math (1929).

b.   The case $K = 1$ of Wiener's Tauberian theorem corresponds to the sequence $k_0 = 1$, $k_n = \infty$ for $n \geqslant 1$ and to the class $C\{k_n\}$ of all bounded infinitely differentiable functions.   Clearly, this class is not quasi-analytic and the integral (4.4) converges.

Remark 4.2.   Theorem 4.2 has been extended in a profound way by Nyman, Domar, and Korenbljum, cf., [8].   We state a definition and result by Domar [9].   A Beurling convolution algebra $L_K^1(\mathbb{R})$ is a __spectral algebra__ if the spectral analysis question has an affirmative answer.   The theorem is that spectral algebras $L_K^1(\mathbb{R})$ are characterized by the property that

$$\text{for all} \quad \gamma \in \mathbb{R}, \quad \Sigma \frac{\log K(n\gamma)}{n^2} < \infty,$$

and this is equivalent to (4.3).

Remark 4.3.   The __Bernstein approximation problem__ was solved by Harry Pollard (1955) with the following theorem.   Let F be continuous and positive on $\mathbb{R}$ and assume

$$\lim_{|\gamma| \to \infty} \gamma^n / F(\gamma) = 0 \quad \text{for all} \quad n \geqslant 0;$$

the span of $\{\gamma^n / F(\gamma) : n \geqslant 0\}$ is sup-norm dense in $C_0(\mathbb{R})$ if and only if

$$\sup_{|K|<F} \int_{-\infty}^{\infty} \frac{\log^+|K(\gamma)|}{1 + \gamma^2} d\gamma = \infty$$

($\log^+ y = \max\{0, \log y\}$, $y > 0$), where K is a real polynomial, cf., G. Wahde, Math. Scand. 20 (1967), 209-224. We mention this result, not only because of its relation to Kolmogoroff's Theorem 4.1 (with the divergent integral) and the spectral analysis question for quasi-analytic classes, but because Pollard's basic lemma on entire functions of exponential type was used by de Branges in dealing with the likes of Theorem 1.3.

5.   DURATION HYPOTHESES; SAMPLING, UNCERTAINTY, AND ENTROPY

We close with a few brief remarks on sampling, uncertainty, and entropy: sampling can be construed as a uniqueness theory in the sprit of Section 1; the uncertainty principle is often invoked to establish weaknesses in periodogram spectrum estimation; and a proper understanding of entropy rate for a given spectrum estimation problem is essential to implement Theorem 3.1 (MEM) correctly.

Let us begin with the Shannon sampling theorem, cf., the treatment and historical remarks in [25] as well as the work of H. Landau, e.g., [19].

Theorem 5.1.   Let $\hat{f} = F \varepsilon L^1(\mathbb{R})$ and assume supp $F \subseteq [\Omega_0, \Omega_0 + 2\Omega]$.

a.   If $T < 1/\Omega$ then for all $t \varepsilon \mathbb{R}$,

$$f(t) = \Sigma \, f(nT)\frac{\sin 2\pi(t-nT)\Omega}{2\pi(t-nT)\Omega} \qquad (5.1)$$

b.   If $T < 1/\Omega$ and $f(t_0 + nT) = 0$ for all n then $f = 0$.

The following "one-line" proof (5.2) is due to Harry Pollard and Shisha (1972):

$$f(t) = \int_{-\Omega}^{\Omega} (\Sigma \, a(n) e^{-2\pi i n \gamma / \Omega}) e^{2\pi i \gamma t} \, d\gamma \qquad (5.2)$$

$$= \Sigma \, a(n) \int_{-\Omega}^{\Omega} e^{2\pi i \gamma (t - n\Omega)} d\gamma.$$

The interchange of order of summation and integration follows since the product of Fourier series and functions of bounded variation can be integrated term by term.

Remark 5.1. If $T > 1/\Omega$ we have aliasing, i.e., the periodic repetitions of F on $\mathbb{R}$ which occur by treating the Fourier series of F (the middle term of (5.2)) will overlap and so the high frequecies will assume the "alias" of the low frequencies.

Suppose that instead of points $\{t_0 + nT\}$ we are given an arbitrary discrete set $D \subseteq \mathbb{R}$. The closure problem is to find the upper bound Ω of the set of numbers ω for which the span E of $\{\exp(2\pi i \gamma t) : t \, \varepsilon \, D\}$ is dense in $L^2[-\omega, \omega]$. Beurling and Malliavin [7] solved this problem completely by writing Ω in terms of a density condition on D. The condition is complicated and the proof is more so; once again, the behavior of integrals (1) plays a role.

Let us point out why the condition $\bar{E} = L^2[-\Omega, \Omega]$ is a sampling theorem. Let $F \, \varepsilon \, L^2(\mathbb{R})$ be supported by $[-\Omega, \Omega]$. Because of the condition $\bar{E} = L^2[-\Omega, \Omega]$ we can find a sequence of polynomials $P_n(\gamma)$ with frequencies $D_n \subseteq D$ such that

$$\lim_{n \to \infty} \| F - P_n \|_{L^2[-\Omega, \Omega]} = 0.$$

If $\hat{p}_n = P_n$ then $p_n = \Sigma \, a_t \delta_t$, $t \, \varepsilon \, D_n$. Thus, for $\hat{f} = F$ we have

$$\| F - P_n \|_{L^2[-\Omega, \Omega]} = \| F - P_n \chi_\Omega \|_{L^2(\mathbb{R})} = \| f - p_n * d_\Omega \|_{L^2(\mathbb{R})},$$

where d_Ω is the Dirichlet kernel $(\hat{d}_\Omega = \chi_\Omega)$. Consequently,

$$\lim_n p_n * d_\Omega = f$$

in $L^2(\mathbb{R})$, and this is precisely the form of (5.1) since $p_n * d_\Omega(t) = \Sigma \, a_u d_\Omega(t - u)$, $u \, \varepsilon \, D_n$.

A distinction between the uniquenss theory of Section 1 and that of the sampling theorem is that the former can be considered a <u>time gap</u> result whereas the latter depends on <u>band-limited</u> hypotheses.

The uncertainty principle and some aspects of entropy have already been discussed in [25]. In light of the differences between periodogram spectrum estimation and MEM spectrum estimation, and the fact that the uncertainty principle plays a role in the former and entropy rate in the latter, we close by recalling a connection between uncertainty and entropy inequalities.

In a 1924 Göttingen seminar, Wiener established the famous and simple (uncertainty principle) inequality,

$$\| t f(t) \|_2 \| \gamma F(\gamma) \|_2 > 1/4\pi, \tag{5.3}$$

for $\hat{f} = F$ and $\| f \|_2 = 1$.

Given a probability space (P, p) and a real random variable $x : P \to \mathbb{R}$. Let p_x, a probability measure on $\mathbb{R}$, be the distribution of x on $\mathbb{R}$ and let $F_x(\gamma) = p\{\alpha \, \varepsilon \, P : x(\alpha) < \gamma\}$ be its distribution function so that $F_x' = p_x$ distributionally, in the sense of Schwartz. The <u>entropy</u> of p_x is

$$H(F_x) = - \int_{-\infty}^{\infty} F_x(\gamma) \log F_x(\gamma) d\gamma,$$

and, using this definition, we say, loosely speaking, that the <u>entropy</u> of F is $H(F) = -\int_{-\infty}^{\infty} F(\gamma) \log F(\gamma) d\gamma$.

Hirschman [14] proved $H(|f|^2) + H(|F|^2) > 0$ for $\hat{f} = F$ and $\| f \|_2 = 1$, and noted the plausibility of the inequality,

$$H(|f|^2) + H(|F|^2) > 1 - \log 2. \tag{5.4}$$

Indeed, (5.4) is valid using Beckner's constant [11]; and, as Hirschman pointed out, (5.4) <u>includes</u> (5.3) by means of an argument found in [26, pp. 55-57].

<u>Example 5.1.</u> In light of the entropy rate used in Theorem 3.1 we point out that $H(F) \leqslant - \int \log F(\gamma) d\gamma$ for $F > 0$.

BIBLIOGRAPHY

1. N. Achiezer, "Theory of approximations", F. Ungar Publishing Co., N.Y. (1956).

2. J. Benedetto, Some mathematical methods for spectrum estimation, (this volume).

3. M. Benedicks, The support of functions and distributions with a spectral gap, <u>Math. Scand.</u> (to appear).

4. A. Beurling, "Sur les integrales de Fourier absolument convergentes et leur applications a une transformation fonctionelle", <u>9th Congres Math. Scan.</u> (1938), 345-366.

5. A. Beurling, On quasianalyticity and general distributions (multigraphed lectures), Stanford University Summer Institute, 1961.

6. A. Beurling and P. Malliavin, On Fourier transforms of measures with compact support, <u>Acta Math.</u> 107 (1962), 291-309.

7. A. Beurling and P. Malliavan, On the closure of characters and the zeros of entire functions, <u>Acta Math.</u> 118 (1967), 79-93.

8. H.G. Dales and W.K. Hayman, Esterle's proof of the Tauberian theorem for Beurling algebras, <u>Ann. Inst. Fourier,</u> Grenoble 31 (1981), 141-150.

9. Y. Domar, Harmonic analysis based on certain commutative Banach algerbas, <u>Acta Math.</u> 96 (1956), 1-66.

10. H. Dym and H. McKean, "Gaussian processes, function theory, and the inverse spectral problem", Academic Press, N.Y. (1976).

11. W. Faris, Inequalities and uncertainty principles, J. Math. Physics 19 (1978), 461–466.

12. U. Grenander and G. Szegö, "Toeplitz forms and their applications", University of California Press (1958).

13. H. Helson and G. Szegö, A probelm in prediction theory, Ann. Mat. Pura Appl. 51 (1960) 107–138.

14. I.I. Hirschman, A note on entropy, Amer. J. Math. 79 (1957), 152–156.

15. K. Hoffman, "Banach spaces of analytic functions", Prentice-Hall, Englewood Cliffs, N.J. (1962).

16. T. Kailath, editor, "Linear least-squares estimation", Benchmark Papers in Electrical Engineering, Dowden, Hutchinson, and Ross, 17 (1977).

17. P. Koosis, Harmonic estimation in certain split regions and a theorem of Beurling and Malliavin, Acta Math. 142 (1979), 275–304.

18. J. Lamperti, "Stochastic processes" Springer-Verlag, N.Y., 1977.

19. H. Landau, Necessary density conditions for sampling and interpolation of certain entire functions, Acta Math. 117 (1967), 37–52.

20. N. Levinson, "Gap and density theorems", AMS Colloquium Publications, 26 (1940).

21. P. Malliavin, On the multiplier theorem for Fourier transforms of measures with compact support, Ark. för Mat. 17 (1979), 69–81.

22. P. Masani and N. Wiener, The prediction theory of multivariate stochastic processes, I, Acta Math. 98 (1957), 111–203.

23.   R.E.A.C. Paley and N. Wiener, "Fourier transforms in the
      complex domain", AMS Colloquium Publications, 19 (1934).

24.   A. Papoulis, "Signal analysis", McGraw-Hill, N.Y., 1977.

25.   J. Price, Uncertainty principles and sampling theorems,
      (this volume).

26.   C. Shannon and W. Weaver, "A mathematical theory of com-
      munication", University of Illinois Press, (1949).

27.   N. Wiener, Oeuvres Volume III, edited by P. Masani, MIT
      Press.

HARMONIC SYNTHESIS - THEORETICAL BOUNDS

Gavin Brown

University of New South Wales
Kensington, N.S.W. 2033
Australia

1. INTRODUCTION

At its simplest harmonic synthesis amounts to combining a finite number of pure vibrations - in other words constructing a trigonometric polynomial. Simple though it sounds, there is active research in the area and I will discuss four examples:

I. There exists $\varepsilon_n \to 0$ and phases $\phi_k = \phi_k(n)$ such that

$$(1-\varepsilon_n)\sqrt{n} \leq |\sum_{k=1}^{n} e^{ik\theta} e^{i\phi_k}| \leq (1+\varepsilon_n)\sqrt{n},$$

for all real θ.
(Byrnes, Körner, Kahane; 1977, 1980)

II. Given decreasing positive A_k with A_k^{-1} concave there are phases $\phi_k = \phi_k(m,n)$ such that

$$\sup_{\theta} |\sum_{n}^{m} A_k e^{ik\theta} e^{i\phi_k}| \leq C(\sum_{n}^{m} A_k^2)^{1/2},$$

where C is an absolute constant and the last sum is taken to be less than one
(Brown, Hewitt; 1978)

171

III. Let X be a finite set of integers then

$$\int \Big| \sum_{k \in X} e^{ik\theta} \Big| d\theta \geqslant C \log\#(X),$$

where C is an absolute constant.
(McGehee, Pigno, Smith; 1982)

IV. Let A_k be decreasing positive and

$$A_{2k} \leqslant \frac{2k}{2k+1} A_{2k-1}.$$

Then, for all positive integers N,

$$A_0 + A_1\cos\theta + A_2\cos 2\theta + \ldots + A_N\cos N\theta > 0,$$

for $0 \leqslant \theta < \pi$.
(Brown, Hewitt; 1983)

There is, of course, no time for an adequate discussion
so I will opt for developing a common theme – don't shade
your eyes, hypothesize, synthesize, majorize!

2. RANDOM COEFFICIENTS

The problem posed by Littlewood, [10], in 1966 was to
show there exist constants C_1, C_2 such that, for N sufficiently large, we may choose complex numbers A_k of unit
modulus to give

$$C_1 N^{1/2} \leqslant \Big| \sum_{k=1}^{N} A_k e^{ik\theta} \Big| \leqslant C_2 N^{1/2}. \tag{*}$$

Byrnes almost solved the problem in [5] and Körner completed
it in [9]. (The reader is urged to consult Kahane's elegant
(but sparingly written account in [8].   The ideas are placed
in an appropriate setting and many interesting consequences,
including (I), are derived by elementary Fourier techniques)
It is only Körner's idea that I want to discuss. Accordingly
it is convenient (and close to the truth!) to attribute to

Byrnes the proof of (*) subject to a relaxed hypothesis that certain of the A_k may have modulus strictly less than one. We make this precise by saying that this should be true of the first M coefficients and the last $M/4$ coefficients, where the order of magnitude of M is $N^{3/4}$. The full result can then be obtained using the following.

Lemma 1.   Given $P(\theta) = \sum_{k=L+1}^{L+M} P_k e^{ik\theta}$, with $|P_k| \leqslant 1$, there

exists $Q(\theta) = \sum_{k=L+1}^{L+M} Q_k e^{ik\theta}$, with $|Q_k| = 1$, and

$$\|P - Q\|_\infty = \sup_\theta |P(\theta) - Q(\theta)| \leqslant 5\sqrt{M \log M}.$$

Assuming for the moment that the lemma is true, we may obviously use it to alter A_k, $k = 1,\ldots,M$ and $k = N-M+1, \ldots,N$ with a total error of

$$10\sqrt{M \log M} ,$$

a quantity which is $O(N^{1/2})$. Thus the Littlewood result will certainly follow from Byrnes' result, provided we establish Lemma 1. What is interesting about Lemma 1 is that it follows by a simple (but ingenious) random choice. It is an easy exercise to deduce the lemma from another which we now state.

Lemma 2.   Suppose that R_k, $k = 1,\ldots,M$, are independent

random variables with $|R_k| \leqslant 1$, and let $R(\theta) = \sum_{k=L+1}^{L+M} R_k e^{ik\theta}$.
Then

$$\text{Prob}\{\|R\|_\infty > \lambda \sqrt{M \log M}\} \leqslant 4M^2 M^{-\lambda^2/5}.$$

Assume Lemma 2.   Given $|P_k| \leqslant 1$, we choose $\pm S_k$ to be the two points of intersection of the circles with centres $\pm P_k$ and radius 1. (Thus $|P_k \pm S_k| = 1$) We let $R_k = +S_k$ with probability 1/2, $-S_k$ with probability 1/2. Taking

$M > 2$, $\lambda = 5$, we have

$$\text{Prob}\{\|R\|_\infty > \lambda \sqrt{M\log M}\} < 1 \,,$$

so the complementary event, $\{\|R\|_\infty < \lambda \sqrt{M\log M}\}$ must occur. Finally we obtain the result of Lemma 1 by choosing $Q_k = P_k - R_k$.

Now let me show how to obtain Lemma 2 starting from the following variation on the theme of the Central Limit Theorem.

Lemma 3.   Suppose that X_k, $k = 1,\ldots,M$, are independent complex random variables of modulus not exceeding one. Let $S = \sum_{k=1}^{M} X_k$, then

$$\text{Prob}\{ |S-ES| > \lambda \} < 4e^{-\lambda^2/4M}.$$

Proof of Lemma 3.   We suppose that $0 < X_j < 1$ and let $Y_j = X_j - EX_j$ so that $|Y_j| < 1$. For a parameter u, $0 < u < 1$, we have

$$E(\exp u\, Y_j) = 1 + \frac{u^2}{2} E(Y_j^2) + \sum_{n=3}^{\infty} \frac{u^n}{n!} E(Y_j^n)$$

$$< 1 + u^2(\frac{1}{2!} + \frac{1}{3!} + \ldots)$$

$$< 1 + u^2 < \exp(u^2).$$

Thus

$$E(u(S-ES)) = \prod_{j=1}^{M} E(\exp(uY_j)) < \exp(Mu^2).$$

$$\text{Prob}\{S\text{-}ES > (t+Mu^2)/u\} \leqslant \text{Prob}\{\exp(uS\text{-}uES) \geqslant e^t E(\exp(uS\text{-}uES))\}$$

$$\leqslant e^{-t}.$$

Now take $t = \lambda^2/4M$, $u = \lambda/2M$ to get

$$\text{Prob}\{S\text{-}ES \geqslant (\lambda^2/4m + \lambda^2/4M) \times (2M/\lambda)\} \leqslant e^{-\lambda^2/4M},$$

provided $\lambda \leqslant 2M$. Splitting the general X_j into real and imaginary parts, positive and negative parts, we find that, for $\lambda \leqslant 2M$,

$$\text{Prob}\{S\text{-}ES \geqslant \lambda\} \leqslant 4e^{-\lambda^2/2M}.$$

The case where $\lambda \geqslant 2M$ is much easier. In fact we have $|S\text{-}ES|^2 \leqslant M$, $\exp|S\text{-}ES|^2 \leqslant \exp M$, where, as previously, we take positive random variables. Then

$$\text{Prob}\{|S\text{-}ES| \geqslant \lambda\} \leqslant \text{Prob}\{\exp|S\text{-}ES|^2 \geqslant \exp\lambda^2\}$$

$$\leqslant \text{Prob}\{\exp|S\text{-}ES|^2 \geqslant \exp(\lambda^2/4M)E(\exp|S\text{-}ES|^2)\}$$

$$\leqslant \exp(-\lambda^2/4M).$$

(since $\exp(\lambda^2/4M)\exp M \leqslant \exp(\lambda^2)$, for $\lambda \geqslant 2M$). For complex random variables we find once more

$$\text{Prob}\{|S\text{-}ES| \geqslant \lambda\} \leqslant 4\exp(-\lambda^2/4M).$$

<u>Proof</u> that <u>Lemma 2</u> <u>follows</u> <u>from</u> <u>Lemma 3</u>.   The idea is to use a grid of points,

$$\theta_j = 2\pi j/M^2, \quad j = 1, 2, \ldots, M^2.$$

We have that every point θ in $[0, 2\pi]$ is within $2\pi/M^2$ of some θ_j and hence

$$|R(\theta) - R(\theta_j)| \leq \sum_{k=1}^{M} 2\pi k/M^2 < 2\pi.$$

Given λ, M we choose α such that $M^\alpha \geq 10\pi/\lambda$. It follows that

$$|R(\theta)| \leq |R(\theta_j)| + \lambda M^\alpha/5,$$

and hence that

$$\text{Prob}\{\|R\|_\infty \geq \lambda M^\alpha\} \leq \text{Prob}\{|R(\theta_j)| \geq 4\lambda M^\alpha/5, \qquad j=1,\ldots,M^2\}$$

$$\leq 4M^2 \exp(-\lambda^2 M^{2\alpha-1}/5).$$

In other words,

$$\text{Prob}\{\|R\|_\infty \geq \mu\} \leq 4M^2 \exp(-\mu^2/5M),$$

and the required result follows when we choose $\mu = \lambda\sqrt{M\log M}$.

3. NON-RANDOM COEFFICIENTS

We turn to (II). The problem arose in a study of Fourier transforms of singular measures. The inequality between the L^∞ and L^2 norms is, of course, the "wrong way round" and typical constraints on the A_k were that A_k should tend to zero as quickly as is compatible with divergence of

$$\sum_{k=1}^{\infty} A_k^2.$$

(Standard examples are $A_k = k^{-1/2}, A_k = (k\log k)^{-1/2},\ldots,$ $A_k = (k\log k\log\log k\ldots\log_{(s)}k)^{-1/2}.$) The best result which can be obtained (in any natural way) by randomizing the phases

would make it necessary to replace the constant C by a term
of order of magnitude $\sqrt{\log(m-n)}$, and this would have been
useless for the applications in view.

Fortunately Salem, [14] has provided methods (based on
the van der Corput lemmas) for making specific phase choices
in related situations and we were able to exploit this in
[2], [3]. Here is our recipe for choosing ϕ_k:

First interpolate a nonincreasing continuously differen-
tiable function $f(x)$ for $n \leqslant x \leqslant m$, such that $f(k) = A_k$
and $1/f$ is concave. Let

$$\phi_k = 2\pi \int_n^k \int_n^y f(x)^2 dx dy / \int_n^m f(x)^2 dx.$$

In the special case where $A_k \equiv 1/M$, $n = 1$, $m = M+1$ this
gives the polynomial space inequality

$$\sup_\theta \left| \sum_{k=1}^{M+1} \exp\left(\frac{\pi i(k-1)^2}{M^2} \right) \exp(ik\theta) \right| \leqslant C \sqrt{M},$$

which links this section with the previous one.

4.   THE LITTLEWOOD CONJECTURE

The question of the truth of (III) has been around since
1948 (cf. [7]) and has generated much interesting mathematics
as by-products of unsuccessful attempts to solve it (see [6],
[12]).   It turns out to be much easier than expected – a
fact which reflects great credit on those who defied conventional wisdom in finding the solution! We can readily view
the problem as another exercise where we must make the uni-
form norm $\|T\|_\infty$ of a trigonometric polynomial small while
keeping the coefficients $\hat{T}(n)$ large.   In fact, for f in
L^1 we have

$$f*T(0) = \Sigma \ \hat{f}(n)\hat{T}(n),$$

and hence, for $\|f\|_\infty \leqslant 1$,

$$\|f\|_1 \geq \|f\|_1 \|T\|_\infty \geq |\Sigma \hat{f}(n)\hat{T}(n)|.$$

Let us enumerate $X = (n_j)_{j=1}^M$ and let $f(\theta) = \underset{n\varepsilon X}{\Sigma} e^{in\theta}$.
Then

$$\|f\|_1 \geq |\overset{M}{\underset{j=1}{\Sigma}} \hat{T}(n_j)|$$

so it will suffice to find a trigonometric polynomial T which satisfies

i. $\|T\|_\infty \leq 1$

ii. $|\hat{T}(n_j)| \geq C\,j^{-1}, \quad j = 1,\ldots,M$

iii. $|\text{sgn}\,\hat{T}(n_j)-1| < \frac{1}{2}$, say.

(Condition (iii) ensures that $|\Sigma \hat{T}(n_j)| \geq 1/2\,\Sigma\,|\hat{T}(n_j)|$ and condition (ii) shows this is greater than $(1/2)C \log M$).

With the Littlewood conjecture expressed as a majorization problem of the type considered in §§1,2 one might be tempted to search for some subtle arithmetical pattern for the coefficient array. The successful proof, given in [11], is much more workmanlike. T is built using "obvious" functions of the form,

$$f_n(\theta) = \#(X_n)^{-1} \underset{j\varepsilon X_n}{\Sigma} e^{ij\theta},$$

where $(X_n)_{n=1}^N$ is a suitable partition of X. The ingenuity comes in finding a clever way to combine the f_n inductively to produce T.

5. POSITIVE SUMS

In result (IV) one settles for one-sided bounds and can therefore make do with a very weak hypothesis on the amplitudes. (Yet again hypothesize, synthesize, majorize.) Two classical special cases are

$$1 + \cos\theta + \frac{1}{2}\cos 2\theta + \frac{1}{3}\cos 3\theta + \ldots + \frac{1}{N}\cos N\theta > 0, \quad (\text{Young},[16]),$$

and

$$\frac{1}{2} + \frac{1}{2}\cos\theta + \frac{1}{3}\cos 2\theta + \frac{1}{4}\cos 3\theta + \ldots + \frac{1}{N+1}\cos N\theta > 0,$$

(Rogosinski, Szegö [13]).

A profound generalization of Young's inequality which does not however contain the Rogosinski-Szegö inequality was given by Vietoris [15] and corresponds to the hypothesis

$$A_{2k} \leqslant \frac{2k-1}{2k} A_{2k-1}.$$

(IV) appears to be very much more difficult to prove, although some parts of the proof follow Vietoris's guidelines. The corresponding sine series result is true for odd N and false for even N (The only bad sub-interval being contained in $[\pi-\pi/N,\pi]$, it is still possible to obtain known results for positivity of sine series with minimum effort).

The first step in the proof is to use summation by parts to reduce to the case where the amplitudes A_k are given explicitly by $A_k = d_k$, and

$$d_{2k} = d_{2k+1} = 2^{2k}/(k+1)\binom{2k+1}{k}$$

Improbable though it may seem at first glance it is easy to sum the infinite series

$$\sum_{k=0}^{\infty} d_k \cos k\theta$$

explicitly. Because the sum to N terms exceeds the quantity

$$\sum_{k=0}^{\infty} d_k \cos k\theta - d_{N+1} \operatorname{cosec}\frac{1}{2}\theta,$$

it is possible to combine estimates for the infinite sum with asymptotic estimates for d_N to prove that the Nth partial sum is positive in the interval $[2\pi/N, \ \pi/19]$ for $N > 38$. Another elementary technique (this time borrowed from Vietoris) gives a way of showing that <u>all</u> the cosine polynomials are positive on $[\pi/19, \pi[$. (Although elementary, this step is rather long, and involves, for example, locating the roots of a certain polynomial of degree 18 whose coefficients are of order 10^9). It is simple to prove the result for $0 < \theta < \pi/N$, so we are left with a gap from π/N to $2\pi/N$. It is on the interval $[\pi/N, 2\pi/N]$ that some genuinely new idea is required — but old ideas are best and we were fortunately able to develop an <u>ad hoc</u> analogue of classical techniques which exploit convexity via partial summation! The relevant (much attenuated) form of convexity turns out to be the following property of the special sequence (d_k):

$$\sum_{r=0}^{s} (d_r - d_{2m-r} - d_{2m+r+1}) > 0, \quad s = 0,1,\ldots,m-1.$$

In relating (IV) to earlier sections, one should note that the order of magnitude of d_k is $k^{-1/2}$. This makes plausible an application to the singular measure question discussed in [2]. It would be interesting to have majorization results in which the amplitudes decrease less rapidly.

Askey and Steinig [1] showed <u>inter alia</u> how to embed the Vietoris result in a wider special function context. Professor Askey has recently informed Hewitt and myself of an appropriate way of applying the main result of [4] (at least for odd N) to the study of Jacobi polynomials.

REFERENCES

1. R. Askey and J. Steinig, Some positive trigonometric sums, <u>Trans</u>. <u>Amer</u>. <u>Math</u>. <u>Soc</u>. 187(1974), 295-307.

2. G. Brown and E. Hewitt, Some new singular Fourier-Stieltjes series, <u>Proc</u>. <u>Nat</u>. <u>Acad</u>. <u>Sci</u>. <u>USA</u> 75(1978), 5268-5269.

3. G. Brown and E. Hewitt, Continuous singular measures with small Fourier-Stieltje transforms, <u>Advances</u> <u>in</u> <u>Math</u>. 37(1980), 27-60.

4. G. Brown and E. Hewitt, A class of positive tri-
    gonometric sums, Math. Ann. 268 (1984), 91–122.

5. J.S. Byrnes, On polynomials with coefficients of modulus
    one, Bull. London Math. Soc. 9(1977), 171–176.

6. P.J. Cohen, On a conjecture of Littlewood and indempo-
    tent measures, Amer. J. Math. 82(1960), 191–212.

7. G.H. Hardy and J.E. Littlewood, A new proof of a theorem
    on rearrangements, J. London Math. Soc. 23(1948), 163–
    168.

8. J.-P. Kahane, Sur les polynomes a coefficients unimodu-
    laires, Bull. London Math. Soc. 12(1980), 321–342.

9. T. Körner, On a polynomial of S.J. Byrnes, Bull. London
    Math. Soc. 12(1980), 219–224.

10. J.E. Littlewood, On polynomials $\Sigma \pm z^m$, $\Sigma \ e^{\alpha_m i} z^m$, $z = e^{\theta i}$,
    J. London Math. Soc. 41(1966), 367–376.

11. O.C. McGehee, L. Pigno, and B.P. Smith, Hardy's inequal-
    ity and the L^1 norm of exponential sums, Ann. of Math.,
    113(1981), 613–618.

12. S.K. Pichorides, On the L^1 norm of exponential sums,
    Ann. Inst. Fourier, 30(1980), 79–89.

13. W. Rogosinski and G. Szegö, Uber die Abschnitte von
    Potenzreihen, die in einem Kreise beschrankt bleiben,
    Math. Z. 28(1928), 73–94.

14. R. Salem, Generalisation de certaines lemmes de Van der
    Corput et applications aux series trigonometriques C.R.
    Acad. Sci. Paris 201(1935), 470–472.

15. L. Vietoris, Uber das Vorzeichen gewisser tri-
    gonometrischer Summen, S.-B. Öster. Akad. Wiss.
    167(1958), 125–135 Teil II Anzeiger Öster. Akad. Wiss.
    1959, 192–193.

16. W.H. Young, On a certain series of Fourier, Proc. London
    Math. Soc. 11 (1912), 357–366.

SEISMIC IMAGE PROCESSING FOR PETROLEUM EXPLORATION

P. W. Buchen

Department of Applied Mathematics
University of Sydney
Sydney, Australia

The high cost of drilling (that is, direct sensing) has made seismic profiling (that is, remote sensing) the most important single tool in the exploration for oil and gas in the sedimentary basins of the Earth's crust. Vast quantities of data are collected at the fraction of the cost of a single well. These digital data are subjected to computer processing in order to extract from them interpretable information of the sub-surface. At the very basic level, the processing yields an acoustic reflectivity map. Colour image processing can enhance geologically significant features such as rock type, porosity, age, lithology, fluid content and structure.

This paper attempts to introduce to the non-specialist some of the basic concepts in seismic data processing which lead to the construction of such interpretable reflectivity maps. Techniques of velocity filtering, deconvolution and migration will be reviewed. New developments in image display and effects of 3-dimensionality will be discussed.

FOURIER THEORY IN MODERN IMAGING

T.W. Cole

Laboratory for Imaging Science and Engineering
School of Electrical Engineering
University of Sydney

SUMMARY

Fourier theory is a powerful tool to the understanding
of imaging systems. But imaging has evolved in a number of
ways and the theory of these imaging systems has developed in
a way far removed from the classical optical system. Much of
the innovation arose in radioastronomy where the millionfold
longer wavelength than in optics meant new approaches were
needed before resolution matched the optical case.
Radiotelescopes now surpass optical telescopes in resolution
by the use of new techniques. These techniques illustrate
the limitations of Fourier approaches. There are strong
parallels with medical imaging where image synthesis, iterative processing and other digital techniques are used in
place of a direct Fourier approach to image generation.

INTRODUCTION

The radioastronomy image is an optical representation of
the measured radio emission across the sky. Such emission
can be measured as a function of frequency, polarisation,
position, and time. The breadth of phenomena to be studied
has led to a plethora of imaging types and an attempt is now
made to categorise the instrumentation into a number of
classes in order to see how the imaging theory has evolved
[1], [2].

DIRECT IMAGING

Just as an optical telescope, with its lenses or mir-
rors, can form an image of the distant scene, so too can a
radio analogue of such an instrument but with several important differences.

The parabolic or spherical reflector is a direct analogy
of Newton's reflecting telescope. The reflector focuses the
incident waves to a single receiver at the focal point. A
typical example of such a telescope is the 64m reflector at
Parkes [3]. The radio spectrum does not have the two-
dimensional detectors of optics, (e.g. photography) and an
image is created by point-by-point scanning.

The lens also has an analogue in the radio antenna array
as illustrated in Figure 1. A lens focuses by introducing
delay to rays from different parts of the aperture. The
antenna array can achieve focusing by inserting appropriate
cable lengths between each element and the final summation
point. By variation of the lengths of cable, and thereby the
delays, rays at different angles of incidence can be
focussed. Once again the radio case does not have a two-
dimensional image plane – just a single image point for each
set of cables.

It is an established observation that the operation of a
lens as in Figure 1(a) can be described as a Fourier
transformation [4] and hence the arrangements in (b) and (c)
generate one point of the Fourier transform. One interesting
development here is the use of a matrix of wires intercon-
necting the antenna array elements and having a number of
output connections, each representing the focussed energy
from a different angular direction [5]. Such a "hardwired"
transform can be constructed to have low chromatic aberrations with all the delays exactly those required to achieve
aberration-free imaging. But such a transform is not a
Fourier transform since the Fourier integral is concerned
with phase and not delay. The Fourier integral is appropriate for monochromatic imaging and chromatic aberrations will
become apparent with wide bandwidth observations if only
phase is considered. (The FFT or Fast Fourier Transform has
its equivalent in analogue form. The so-called Butler Matrix
uses cables whose phase lengths at the central operating frequency correspond to the phase rotation factors of the FFT.)

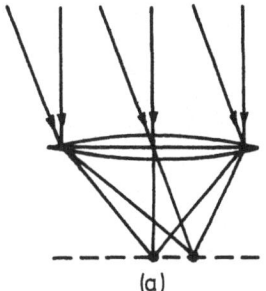 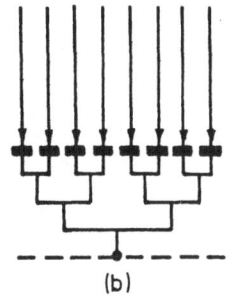 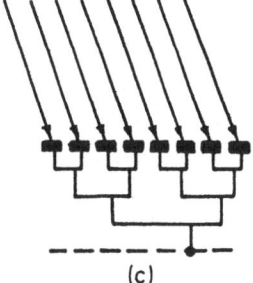

(a)   (b)   (c)

Figure 1

An array of antenna elements can be connected to form an
analog of the optical lens shown in (a). Radiation from
a given direction is delayed so that it adds in phase at
one point of the image plane. The radio case is illustrated in (b) and (c) where it is seen that different
cable arrangements are needed for each image plane
point.

A number of antenna arrays for direct imaging have been
built with a wide range of geometries and complexities. An
extreme application of the approach, and of the direct imaging concepts used, is the Culgoora radioheliograph [6]. The
CSIRO instrument at Culgoora (near Narrabri, N.S.W.) consists
of 96 antenna elements distributed around the circumference
of a 3km diameter circle. The imaging problem is to achieve
with an annular aperture the same imaging performance as a
completely filled aperture 3km in diameter. Such a problem
had not been faced in optical imaging and the circular
geometry of Culgoora represented a significant extension to
the techniques already in use in radioastronomy at that time
which had regular geometries such as crosses or T-shaped
arrays.

IMAGE CORRECTION WITH AN IRREGULAR APERTURE

The Culgoora heliograph is a dilute aperture, with the
consequence that its response to a point source (point spread
function) contains a much higher level of nearby responses or
sidelobes. This leads to confusion when imaging a complex
source. Whereas a circular aperture has as a point spread
function the well-known Airy function ($J_1^2(x)/x^2$), the annular

aperture has a function $J_0^2(x)$ where J_0 and J_1 are Bessel functions. These are illustrated in Figure 2(a) and 2(b). In Figure 2(c) is shown the response of the heliograph where, in addition to the central $J_0^2(x)$ response one sees an outer ring of "grating" responses due to the use of only 96 elements to approximate a continuous annular aperture.

The process by which the heliograph (or any aperture) is able to achieve images with the Airy (or any other) point spread function of the same general resolution is better understood in a Fourier description [7], [8].

In Cartesian co-ordinates the antenna may be specified by its aperture distribution g(u,v) where u and v, measured in wavelengths, denote rectangular co-ordinates in the plane of the aperture. The directional response of the antenna is specified in terms of its amplitude polar diagram, f(1,m), or its power polar diagram, F(1,m), where (1,m) denotes a

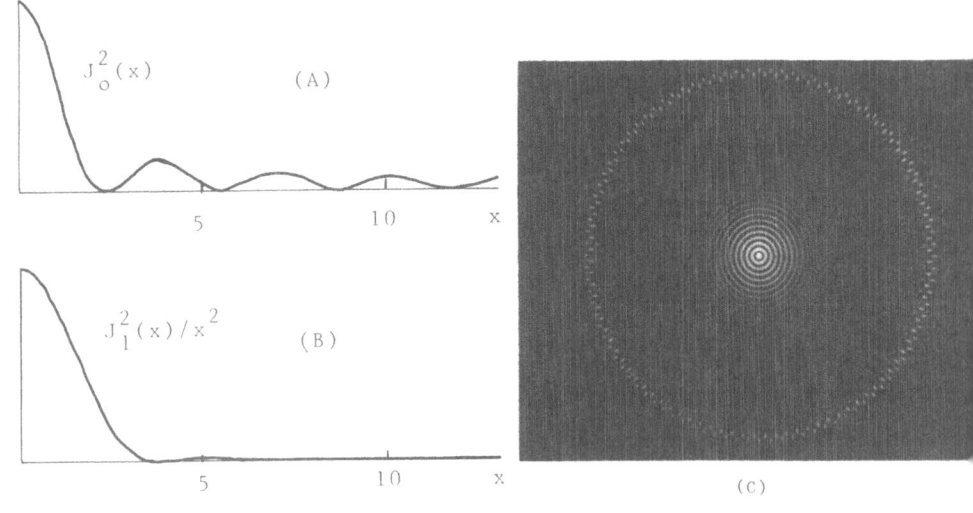

Figure 2

The power polar diagram profile is shown in (a) for the uniformly illuminated circular aperture and in (b) for the continuous ring-shaped aperture. In (c) is shown the two-dimensional response of the heliograph including the grating ring due to the use of only 96 elements around the ring.

rectangular system of angular co-ordinates parallel to (u,v) over a small part of the celestial sphere. Then,

$$F(l,m) = |f(l,m)|^2$$

and

$$f(l,m) = \iint_{-\infty}^{\infty} g(u,v) \exp(2\pi i(ul+vm)) \, du \, dv.$$

This latter equation merely restates that the amplitude polar diagram and the aperture distribution are Fourier pairs.

The Fourier transform of the power polar diagram is referred to as the spatial frequency sensitivity of the antenna and by the standard Fourier theorem, it is also obtainable as the autocorrelation of the aperture distribution.

The heliograph has circular symmetry and these equations reduce [8] to

$$f(r) = 2\pi \int_0^\infty g(m)J_0(2\pi mr)m \, dm$$

where r and m are radial co-ordinates in the aperture and sky planes respectively.

Changing the power polar diagram of the antenna is well known to be possible by modification of its spatial frequency sensitivity [7]. The modification is effected by the suppression of different components in different degrees and quite clearly no missing components can be recovered by these means and no great fundamental increase in resolution is possible for the instrument. However the sidelobes or nearby responses can be modified. The modification is therefore achieved by a filtering operation in the spatial frequency plane, and again by standard theory, it could also be achieved by a convolution process in the directly-recorded image plane by some suitable two-dimensional correcting function [8].

In practice, the heliograph achieves correction by a technique known as J^2 synthesis [9], [6]. This is described in the references but in essence, when different phase gradients are inserted around the annulus, the point spread function can be shown to be $J_n^2(x)$ for a gradient of $2\pi n$ radians around the annulus.

The (intensity) images obtained with these different point spread functions are combined in a linear combination

with (positive and negative) weights. Choice of weights
enables any point spread function to be created within the
resolution limit of the array. The process effects a
reweighting of the spatial frequency components and offers a
new degree of freedom to the imaging compared with classical
optics. It is now possible to create a point spread function
in which some of the nearby responses are negative in intensity. This is physically unrealisable in an optical system
using only masks or filters to weight (apodize) in the aperture plane rather than the spatial frequency plane. The
extra freedom possible by this now not only corrects for
dilute apertures but also enables point spread functions to
be used which have a more optimum compromise between resolution and magnitude (positive or negative) of the confusing
nearby responses. That is, images are now created for which
there is no real aperture corresponding to the point spread
function in these images!

However, this is possible only if the aperture is sensitive to all the spatial frequency components within some
limit determining the resolution. Admittedly the sampling
theory allows a quantised sampling if the object to be imaged
is of limited extent. This would produce gratings outside
the field of view. But the case of missing information –
where the spatial frequency plane is not sampled sufficiently
is an important one and will be returned to below.

THE INTERFEROMETER

The classical Young's two-slit interferometer produces
fringe outputs from a point object [4]. The interferometer
is sensitive to one spatial frequency component of the image.
By selection of the separation and orientation of the slits,
any spatial frequency component in the image can be measured.

A radio interferometer is a modern version of the slit
interferometer with an improvement made possible by the ease
of manipulation of radio signals compared with optical [1].
The signals from the two elements can be delayed by extra
cables and their amplitudes can be multiplied rather than
just added as in the optical case. The result is a sensitive
instrument able to be steered by added delays to any direction. At that direction the delay paths through the two elements are equal and all frequencies in the signal add in
phase. At angles away from this boresight, the delay

difference implies a different phase difference for the different frequencies and chromatic aberrations become important. These are not discussed here nor is the implication of
the theory of coherence.

Give that an interferometer observation measures one
spatial frequency component, it should be possible to build
up an entire image by a set of inerferometer measurements
sufficient to sample the spatial frequency plane [10], [11],
[12].

Indeed this is a feasible approach and has been implemented with a number of antenna arrays. For example, in the
radioheliograph, multiplication of signals from pairs of
antennas can produce an image after a reweighting and
transformation of the measurements [13].

The Fourier concept of the spatial frequency plane
introduces further flexibility in the approach to imaging.

APERTURE SYNTHESIS

When the object to be imaged is stable with time, the
interferometer measurements need not be taken simultaneously.
Aperture synthesis is then the sequential measurement of the
spatial frequency plane sampled by interferometers of changing separation and orientation. Aperture synthesis has provided enormous power and flexibility to imaging in radioastronomy. One of its most powerful forms is earth rotation
aperture synthesis [12].

As viewed from the sky, the projected orientation and
separation of antenna elements on the Earth's surface trace
out an ellipse as the Earth rotates. Pioneered at Cambridge,
an east-west line of antenna pairs can thereby fully sample
the spatial frequency plane over twelve hours as the antenna
elements track the region of sky under study.

Extensions are many and varied. A full east-west line
of antennas is not needed if, on successive days, the separation of elements on the ground are changed and the measurements combined in a computer before transformation.

The VLA antenna in New Mexico, USA [14], [15] uses a
different geometrical arrangement - 27 elements on a Y shaped

set of baselines – and observes for eight hours. Indeed the
VLA raises again the problem of insufficient samples of the
spatial frequency plane. In the VLA, the ellipses traced out
by the 27 × 26 × 0.5 = 351 possible antenna pairs form an
irregular density of ellipses in the spatial frequency plane.
The resulting grating responses spread all over the imaging
plane and can no longer be eliminated by simple reweighting.
Once again, the problem of insufficient data appears – as it
does in many other situations in this form of imaging such
as:

      – In the Fleurs instrument [16] it is not possible to
follows a source for the full 12 hours which produces missing
sectors in the spatial frequency plane.

      – In the Very Long Baseline (VLBI) technique which com-
bines telescopes on different continents, one is limited to
existing sites and so only very sparse samplings are possible
[17].

THE IMAGE PROCESSING PROBLEM

      The steps involved in the processing of an aperture syn-
thesis image using Fourier techniques can be illustrated from
the map shown in Figure 3 obtained on the Westerbork array in
the Netherlands [18]. The spatial frequency plane had been
sampled along concentric ellipses in what is essentially a
polar coordinate system. A complete synthesis would measure
enough ellipses on successive days to effectively fill the
spatial frequency plane but this map is a partial synthesis
with the gaps producing the elliptical grating responses concentric with the source region of interest in the map centre.
At least for this source, the map is unambiguous since the
gratings fall outside the region of interest. When this is
not so, there are still ways to deal with it but these techniques are amongst those discussed below which go outside the
area of Fourier techniques.

      Figure 3 was calculated as a digital transformation of
the measurements after they had been interpolated onto a rectangular array of points in the spatial frequency plane. Not
only does this take a large amount of calculation, it also
introduces problems of aliasing whereby the grating responses
appear folded back from the edges of the map and thereby more
likely to cause confusion. To control this effect, the

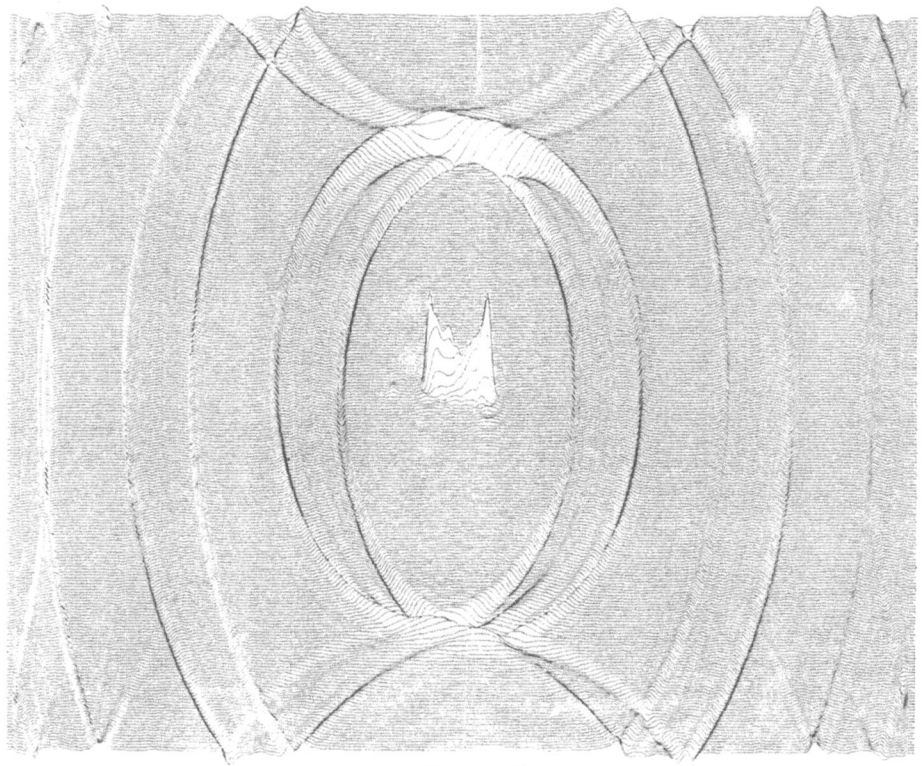

Figure 3

A radio map of source 3C452 taken with the Westerbork
radio telescope illustrates gratings, aliasing, and
sidelobes as well as the source itself in the centre.

elliptical distribution of samples is smeared or convolved
which thereby weights the map down at the edges by a weighting function equal to the Fourier transform of the convolution function. It can also be seen that the source region in
the centre shows evidence of the effect of sidelobes or
nearby ripples. These are the point spread function's ripples which can be controlled by an overall weighting across
the spatial frequency plane.

The process of earth rotation aperture synthesis over-
samples the low-spatial frequency part of the plane since a
12 hour ellipse of samples at low spatial frequencies is
clearly less than at high spatial frequencies. This also

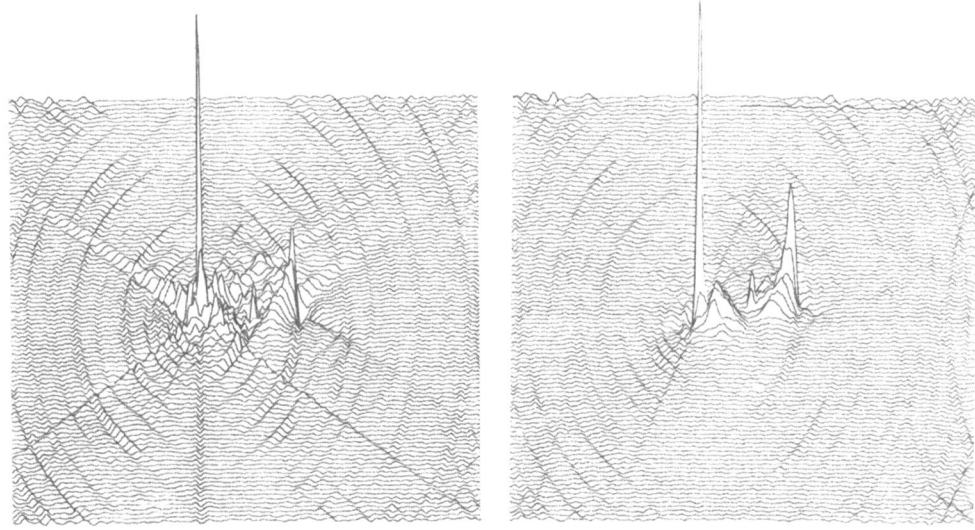

Figure 4

The operation of CLEAN is illustrated by the two images
(a) before, and (b) after, the iterative subtraction of
the grating and sidelobe corrupted point spread function.

needs a correction achieved by another weighting of the spatial frequency plane by a weight proportional to the radius
in the plane. Of course there is an alternative approach to
this correction on the other side of the Fourier transform.
This is the process of ´convolution and back projection´
[19].

MEDICAL IMAGING

It is not appropriate to discourse at length on modern
medical diagnostic imaging [20]. It is appropriate to draw
parallels between the X-ray technique of CT (computerised
tomography) and earth rotation synthesis with an east-west
array of antennas. Both take scans of the image at different
azimuths, process them, and then project back onto the image
plane. Both use identical algorithms and suffer the same
problems of aberrations and gratings from missing or badly
calibrated data. The newer techniques of CT using sector
scans and the even newer techniques of NMR (nuclear magnetic

resonance) imaging are all indirect imaging processes requir-
ing reconstruction from measured data. All are therefore
well removed from the classical optical imaging system,
require extensive computer processing, and can take advantage
of many of the approaches to image processing which go beyond
the capabilities of linear Fourier theory.

ITERATIVE IMAGE RECONSTRUCTION

To overcome the problems of missing information, processing techniques have developed which use a priori information. For example, intensity images are positive. The presence of negative lobes of intensity is representative of the
particular weighting given to the spatial frequency plane,
but also to the grating responses from missing samples.
Iterative approaches to fit an all-positive distribution to
the measurements have shown some success.

A more widely used technique [21], [22] in radioastronomy, called CLEAN, takes advantage of the images usually consisting of bright sources against a dark background. An
iterative subtraction of the (known) point spread function
from the aberrated image results in a distribution of point
source positions which, when replaced by a point-spread function of the same resolution but without gratings, gives an
image remarkably free from the effects of missing samplings
and, in most cases, a true representation of the source.
This is illustrated in Figure 4.

Further correction is possible in some cases by
recognising characteristic shapes in the image which can be
related to some observational error. As seen in Figure 5 the
improvements can be dramatic [23] but do require an intimate
knowledge of the imaging process and the use of iterative and
interactive techniques which go well beyond the Fourier
approaches and their limitations.

More recently there has emerged several new techniques
which are essentially non-linear but which make optimum use
of incomplete or noisy imaging data. Some of the more well
known ones would include Maximum Entropy Methods [24], [25]
and the techniques of phase closure used in the situation
where the phase information from the different interferometers of an array are only partially known [26].

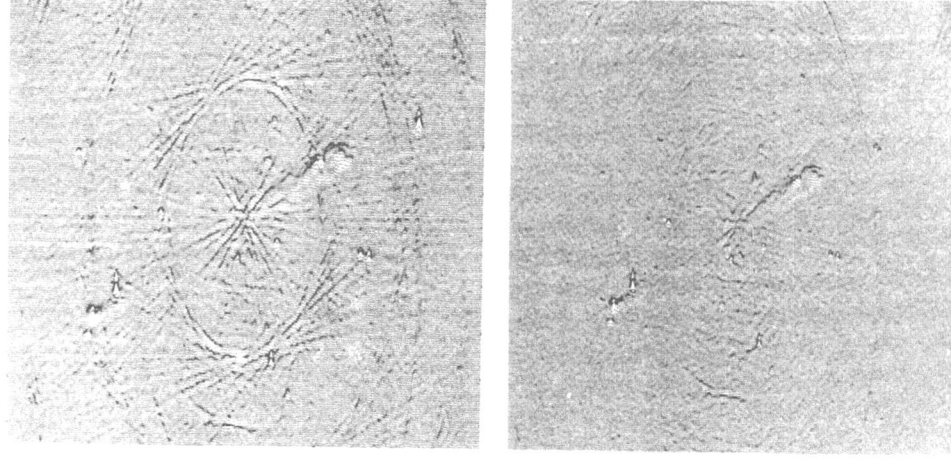

Figure 5

Interactive reduction of residual phase and amplitude
errors. Variable atmospheric effects have produced
radial features around the position of a strong point
radio source which has already been subtracted from the
map. Filtering improves the image to that shown in (b)
where even weaker structure is now clearly visible.

The current capabilities of these imaging techniques and
the computers used to process them extend to images of 2048
by 2048 points and, in addition, up to 256 of such maps might
be produced in each observation period at different radio
frequencies. The quality of the final images can be at the
level that details can be shown whose intensity is 2000 times
less than the intensity of the brightest point in the image.

The field of imaging by these new techniques is advanc-
ing rapidly. It formed the subject of a specialist conference in 1978 in the Netherlands [27]. It was also the topic
of another international conference held in Sydney at exactly
the same time as this conference on the Fourier transform.
The proceedings of that conference [28] represent the best
summary of the current state of these techniques which build
upon the Fourier techniques but which can only achieve their
final, high quality by use of significant advances beyond
them.

REFERENCES

1. W.N. Christiansen and J.A. Hogbom, "Radiotelescopes,"
    Cambridge University press, Cambridge (1969) and (1984).

2. T.W. Cole, Quasi-optical Techniques of Radioastronomy,
    Ch.IV in "Progress in Optices," (Ed. E. Wolf) Vol XV,
    (1977), p.187-244.

3. E.G. Bowen and H.C. Minnett, The Australian 210-foot
    Radio Telescope, Proc. I.R.E.E. Aust. 24(2):98-
    105(1963).

4. M. Born and E. Wolf, "Principles of Optics," Pergamon
    Press, New York (1959).

5. N. Fourikis, The Branching Network, Proc. I.R.E.E. Aust.
    28(9):315-23 (1967).

6. Special Issue on the Radioheliograph, Proc. I.R.E.E.
    Aust. 28 (9) (September, 1967).

7. R.N. Bracewell and J.A. Roberts, Aerial Smoothing in
    Radio Astronomy, Aust. J. Phys. 7(4):615-40 (1954).

8. J.P. Wild, Circular Aerial Arrays for Radio Astronomy,
    Proc. Roy. Soc. A 262:84-99 (1961).

9. J.P. Wild, A new Method of Image Formation with Annular
    Apertures and an Application in Radio Astronomy, Proc.
    Roy. Soc. A 286:499-509 (1965).

10. M. Ryle, Radio Telescopes of Large Resolving Power, Sci-
    ence 188, No. 4193:1071-9 (June, 1975).

11. M. Ryle and A. Hewish, The Synthesis of Large Radio
    Telescopes, Mon. Not. Roy. Astron. Soc. 120 (3):220-30
    (1960).

12. M. Ryle and A.C. Neville, "A Radio Survey of the North
    Polar region, Mon. Not. Roy. Astron. Soc. 125 (1):39-56
    (1962).

13. D.J. McLean, A proposed Correlator Back-end for the Cul-
    goora Radioheliograph, in: "Formation of Images from
    Spatial Coherence Functions in Astronomy," C. van
    Schooneveld, ed., Reidel (1979):159-64.

14. D.S. Heeschen, The Very Large Array, Sky and Telescope
    209:344-51 (1975).

15. D.S. Heeschen, The Very Large Array, in: "Telescopes for
    the 1980s," G. Burbidge and A. Hewitt, eds., Annual
    Reviews Inc., California (1981):1-62.

16. Special Issue, Proc. Inst. Radio. Electron. Eng. Aust.
    34 (8) (September, 1973).

17. P.N. Wilkinson, A.C.S. Readhead, G.H. Purcell and B.
    Anderson, Radio Structure of 3C147 Determined by Multi-
    Element Very long Baseline Interferometry, Nature
    269:764-8 (October 27, 1977).

18. J.A. Hogbom and W.N. Brouw, The Synthesis Radio Tele-
    scope at Westerbork, Astron. and Astrophys. 33:289-301
    (1974).

19. R.N. Bracewell and A.C. Riddle, Inversion of Fan-beam
    Scans in Radioastronomy, Astrophys. J. 150:427-33
    (1967).

20. D. Rosenfeld, Review of Instrumentation in Medical X-ray
    and Gamma-ray Imaging, in:

    Cambridge University Press (1984).

21. J.A. Hogbom, Aperture Synthesis with a Non-regular Dis-
    tribution of Interferometer Baselines, Astron. Astro-
    phys. Suppl. 15:417-23 (1974).

22. U.J. Schwartz, Mathematical-statistical Description of
    the Iterative Beam Removing Technique (Method CLEAN),
    Astron. Astrophys. 65:345-56 (1978).

23. J.P. Hamaker, Kneading, The Adjustment of Instrumental
    Phase and Gain Parameters to Suppress Error Patterns in
    a Synthesis Map in: "Image Formation from Coherence
    Functions in Astronomy," C.van Schooneveld, ed. D.
    Reidel (1979):47-53.

24.   S.J. Wernecke and L.R. d´Addario, Maximum Entropy Image
       Reconstruction, <u>Trans</u>. <u>IEEE</u>, C-26 (4):351-64.(April,
       1977).

25.   S.F. Gull and G.J. Daniell, Image Reconstruction from
       Incomplete and Noisy Data, <u>Nature</u> 272:686-90 (20 April,
       1978).

26.   A.C.S. Readhead, R.C. Walker, T.J. Pearson and M.H.
       Cohen, Mapping Radio Sources with Uncalibrated Visibil-
       ity Data, <u>Nature</u> 285:137-40 (May 15, 1980).

27.   C. van Schooneveld (Editor), "Image Formation from
       Coherence Functions in Astronomy," Proceedings of a
       conference in Groningen, Netherlands, D. Reidel (1979).

28.   J.A. Roberts (Editor), "Indirect Imaging", Proceedings
       of a conference in Sydney, Australia, Cambridge Univer-
       sity Press (1984).

AN OVERVIEW OF TIME AND FREQUENCY LIMITING

H.J. Landau

A.T. & T. Bell Laboratories
Murray Hill, New Jersey
U.S.A.

INTRODUCTION

This note aims to motivate and sketch informally some of the work on time and frequency limiting that I have seen at close quarters. The account is entirely subjective, and is not meant to speak for the friends and collaborators – notably H.O. Pollak, D. Slepian, and B.F. Logan – from whom I learned much of this material, and who could explain it far better. I also apologize to the many other contributors for inadequate mention of their work. What follows is impressionistic and incomplete, intending, as does any brief survey, only to show the interest and charm of the area, with the hope of enticing the reader to return at greater leisure.

A PARADOX

We can begin with the Fourier decomposition. A square-integrable function $f(t)$ – which we term a function of finite energy – has a square-integrable Fourier transform $F(\omega)$, and

$$f(t) = \frac{1}{\sqrt{2\pi}} \int F(\omega)e^{i\omega t} \, d\omega. \tag{1}$$

201

(Unless otherwise stated, all integrals will be over $(-\infty,\infty)$.)
Viewing the integral of (1) as a decomposition into a sum of
frequency components, and believing that physical devices –
vocal chords, membranes, oscillators – all have upper limits
on the rate at which they can vibrate, we can feel confident
that in modeling the outputs of such devices the representation (1) may without loss of generality be corresponding
truncated to

$$f(t) = \frac{1}{\sqrt{2\pi}} \int_{-\Omega}^{\Omega} F(\omega)e^{i\omega t} \, d\omega, \tag{2}$$

for some $\Omega < \infty$, or, equivalently, that

$$F(\omega) \equiv 0, \quad |\omega| > \Omega.$$

We will call square-integrable functions of the form (2)
frequency-limited – or, more specifically, band-limited – and
denote that collection by $B(\Omega)$. Of course, with the same
confidence we can believe that physical devices have zero
response when not activated, so that their outputs may with
equal justice be assumed to have only finite duration

$$f(t) \equiv 0, \quad |t| > T , \tag{3}$$

for some $T < \infty$. We call square-integrable functions of the
form (3) time-limited, and denote that collection by $\underline{D}(T)$.

    The trouble arises because, on extending the variable t
in (2) to the complex $(t+iu)$-plane, we see that a band-
limited function is analytic – indeed, entire – hence cannot
vanish on any interval without vanishing identically. Thus
f(t) cannot be simultaneously band-limited and time-limited.
Another implication of analyticity is extrapolatability: a
band-limited f(t) can be extrapolated everywhere from
knowledge of it on an arbitrarily small interval (say by
means of a power series expansion) so that, for example, any
second´s worth of speech would determine the entire utter-
ance. Clearly, the assumption of band-limitedness seems to
embroil us in paradox. And yet, in the practical design of
communications systems, that assumption, and some of its
theoretical consequences such as sampling, are exploited in
an essential way without generating contradictions. Moreover, according to engineering experience, not only do there

exist signals which are, for all purposes, simultaneously time and band-limited to $|t| < T$ and $|\omega| < \Omega$, respectively, but there are approximately $2\Omega T/\pi$ independent ones. Thus the problem of reconciling the models (2) and (3) presents an interesting challenge [31, 33].

CONCENTRATION AND THE UNCERTAINTY PRINCIPLE

If a band-limited $f(t)$ cannot be supported on a finite interval $|t| < T$, we can nevertheless ask how well it can be concentrated (in energy) on the interval, i.e., how large we can make the ratio

$$\alpha_f^2 = \int_{-T}^{T} |f(t)|^2 dt / \int |f(t)|^2 dt,$$

for $f \in B(\Omega)$. By standard analysis we can conclude that this ratio attains its supremum for a function $\phi_0(t) \in \underline{B}(\Omega)$, that ϕ_0 is the solution of the eigenvalue equation

$$\frac{1}{\pi} \int_{-T}^{T} \frac{\sin\Omega(t-s)}{t - s} \phi_0(s) \, ds = \lambda_0 \phi_0(t), \tag{4}$$

corresponding to the largest eigenvalue λ_0, and that λ_0 is the desired $\sup(\alpha_f^2)$; a change of variable shows $\sup(\alpha_f^2)$; λ_0 to depend only on the product ΩT.

We can profitably interpret this result geometrically, by viewing the space L^2 of square-integrable functions as a Hilbert space, in which

$$(f,g) = \int f(t)\overline{g(t)}dt \, ,$$

$$||f||^2 = \int |f(t)|^2 dt \, ;$$

the Fourier transform is then a unitary operation. We observe that $\underline{D}(T)$ and $\underline{B}(\Omega)$ are closed linear subspaces, and that the (orthogonal) projections of L^2 onto these subspaces are given by the restrictions

$$D_T f = \begin{cases} f(t), & |t| < T \\ 0, & |t| > T, \end{cases}$$

$$B_\Omega f = \frac{1}{\pi} \int \frac{\sin \Omega(t-s)}{t-s} f(s) \; ds \; ,$$

respectively. The expression for B_Ω defines that function whose Fourier transform agrees with $F(\omega)$ in $|\omega| < \Omega$ and vanishes elsewhere; thus the operation B_Ω is completely analogous to D_T, and we refer to it as __frequency-limiting__. In terms of these projections, the concentration of interest becomes $\alpha_f^2 = ||D_T f||^2 / ||f||^2$. Visualizing the projection as in Euclidean space, α_f is the cosine of the angle which f, a vector in $\underline{B}(\Omega)$, forms with the subspace $\underline{D}(T)$; maximizing α_f then corresponds to finding the least angle between the two subspaces. Equation (4) asserts that if this angle is attained at $\phi_0 \; \varepsilon \; \underline{B}(\Omega)$ then

$$B_\Omega D_T \phi_0 = \lambda_0 \phi_0 \; , \tag{5}$$

i.e., that projecting ϕ_0 onto $\underline{D}(T)$ and then back onto $\underline{B}(\Omega)$ yields a vector aligned with ϕ_0. A simple picture in three dimensions will illustrate the geometric reasonableness of this.

The largest eigenvalue of (4) measures the least angle between $\underline{B}(\Omega)$ and $\underline{D}(T)$, and its associated eigenfunction ϕ_0 is the band-limited function best concentrated in $|t| < T$. As such, it and approximations to it have found use in filter design and spectral estimation [10,11,19]. It also plays a role in the study of uncertainty.

The uncertainty principle asserts that a function and its Fourier transform cannot both be concentrated on small intervals: narrowing one necessarily produces broadening of the other. The classical formulation of this considers $|f(t)|^2$ and $|F(\omega)|^2$ normalized as probability distributions,

$$\int |f(t)|^2 dt = \int |F(\omega)|^2 d\omega = 1,$$

and, gauging the spread of each by its variance,

$$\sigma_f^2 = \int t^2 |f(t)|^2 dt$$

$$\sigma_F^2 = \int \omega^2 |F(\omega)|^2 d\omega,$$

establishes that

$$\sigma_f \sigma_F \geqslant 1/2.$$

This result is fundamental in modern physics, where it is interpreted to mean that certain pairs of physical quantities cannot simultaneously be accurately determined. Interesting local versions have recently been proved [25].

With the thought that variance may be a somewhat imprecise measure of dispersion, we can examine uncertainty in terms of energy concentration over specified intervals. Accordingly, with T and Ω fixed, and any $f \in L^2$, let

$$\alpha_f^2 = \int_{-T}^{T} |f(t)|^2 dt / \int |f(t)|^2 dt,$$

$$\beta_f^2 = \int_{-\Omega}^{\Omega} |F(\omega)|^2 d\omega / \int |F(\omega)|^2 d\omega.$$

The set Σ of points (α_f, β_f) in the unit square, generated as f ranges through L^2, describes the possible time and frequency concentrations that are simultaneously attainable. Again, recognizing that $\alpha_f = ||D_T f|| / ||f||$ and $\beta_f = ||B_\Omega f|| / ||f||$, we see that these ratios are the cosines of the angles which f forms with $\underline{D}(T)$ and $\underline{B}(\Omega)$, respectively. It is then a short step to conclude that these angles cannot sum to less than the angle between the subspaces themselves, so that Σ is delimited by the curve $\cos^{-1}\alpha_f + \cos^{-1}\beta_f = \cos^{-1}\lambda_0$ (ellipse inclined at 45°), traced out by linear combinations of ϕ_0 and $D_T\phi_0$. The uncertainty principle reflects itself here in the fact that Σ is bounded away from (1,1) [12].

DOUBLE ORTHOGONALITY OF $\{\phi_k\}$

The problem of maximum concentration has led to the equation (4), which in turn generates other eigenfunctions

$\{\phi_k\}$, $k \geq 0$, and corresponding eigenvalues $\lambda_k = \lambda_k(\Omega T)$. The fundamental properties of an orthogonal projection P are

$$P^2 = P,$$

$$(Pf,g) = (f,Pg), \qquad (6)$$

for any f,g in the Hilbert space. Since $\phi_k \varepsilon \underline{B}(\Omega)$, we can rewrite the operator of (4) and (5) as $B_\Omega D_T B_\Omega$, which shows it to be positive and self-adjoint; it is also compact since the kernel of (4) is square-integrable. It follows that $\{\lambda_k\}$ are positive, approaching 0 as $k \to \infty$, and that $\{\phi_k\}$ are mutually orthogonal functions which, when normalized, form a basis for $\underline{B}(\Omega)$. Moreover, invoking (6),

$$(D_T\phi_i, D_T\phi_j) = (D_T B_\Omega\phi_i, D_T B_\Omega\phi_j) = (B_\Omega D_T^2 B_\Omega\phi_i, \phi_j) =$$

$$(B_\Omega D_T B_\Omega\phi_i, \phi_j) = \lambda_i\delta_{ij}.$$

Thus the eigenfunctions enjoy a remarkable property: not only are they mutually orthogonal over the entire t-axis, but their restrictions to $|t| < T$ are also mutually orthogonal and, when renormalized to $\{\lambda_k^{-1/2} D_T\phi_k\}$, form a basis for $\underline{D}(T)$. This double orthogonality makes $\{\phi_k\}$ an ideal basis for considering the many problems in which information about a square-integrable band-limited function is given on an interval. We illustrate with the question of extrapolation, already mentioned.

Suppose that g(t) is given for $|t| < T$. We ask whether g(t) is the restriction to $|t| < T$ of some $f \varepsilon \underline{B}(\Omega)$ and, if so, what is f? On expanding the given g(t) in the basis $\{\lambda_k^{-1/2} D_T\phi_k\}$, we obtain

$$D_T g = \sum_{k=0}^{\infty} a_k\lambda_k^{-1/2}D_T\phi_k, \qquad (7)$$

where the coefficients a_k are given explicitly by

$$a_k = \int_{-T}^{T} g(t)\phi_k(t)\lambda_k^{-1/2}dt;$$

being components of D_T g in an orthonormal basis, they satisfy

$$\Sigma \ |a_k|^2 = \int_{-T}^{T} \ |g(t)|^2 dt < \infty. \tag{8}$$

The right-hand side of (7), however, is $D_T(\Sigma \ a_k \lambda_k^{-1/2} \phi_k)$. This shows that the desired f must be

$$f(t) = \sum_{k=0}^{\infty} a_k \lambda_k^{-1/2} \phi_k$$

which defines an element of $\underline{B}(\Omega)$ if and only if

$$||f||^2 = \Sigma \ \frac{|a_k|^2}{\lambda_k} < \infty. \tag{9}$$

Since $\lambda_k \to 0$, the extrapolation problem is inherently unstable, or ill-posed in the sense of Hadamard: an insignificant change in $\{a_k\}$ can radically alter the convergence of (9). Of course, qualitatively, this is in no way surprising, since the given data must be analytic to be the restriction of a band-limited function, and analyticity is easily destroyed by arbitrarily small changes. But more specifically, we can see by restricting to finite sums in (7) and (9) that the instability here arises because small improvements in approximating g(t) in $|t| < T$, when contributed by a basis element ϕ_k for which the concentration λ_k is small, can entail arbitrarily large swelling of the size of f(t) away from the interval of observation $|t| < T$. To eliminate the instability we must therefore somehow reduce the influence of these components in (7). One natural way of doing this is to reformulate the extrapolation problem, by asking for the best approximation $D_T f$ to $D_T g$ among functions $f \in \underline{B}(\Omega)$ for which the norm has a fixed bound A, reserving the freedom to vary A. It is easy to see that f here is given by

$$f = \sum_{k=0}^{\infty} \frac{a_k}{\lambda_k^{1/2} (1+\frac{\mu}{\lambda_k})} \phi_k.$$

with μ chosen so that

$$\sum_{k=0}^{\infty} \frac{a_k^2}{\lambda_k (1+\frac{\mu}{\lambda_k})^2} = A^2.$$

A cruder approach might be simply to truncate the expansion (7) of $D_T g$ to those $\{\phi_k\}$ for which λ_k is not too small. To obviate the need for this, R.W. Gerchberg and A. Papoulis [5,5a] suggested the iteration

$$f_1 = B_\Omega D_T g$$

$$f_{n+1} = f_n - BDf_n + BDg, \quad n \geqslant 1.$$

By expanding in terms of $\{\phi_k\}$ it is again easy to see that this procedure generates

$$f = \sum a_k \lambda_k^{-1/2} \phi_k$$

by means of the series

$$\lambda_k^{-1/2} = \frac{\lambda_k^{1/2}}{1-(1-\lambda_k)} = \lambda_k^{1/2}[1+(1-\lambda_k) + (1-\lambda_k)^2 + \ldots],$$

so that

$$f_n = \sum_{k=0}^{\infty} a_k \lambda_k^{-1/2} [1-(1-\lambda_k)^n] \phi_k,$$

a formula in which the weights $[1-(1-\lambda_k)^n]$ attached to the coefficients have, at least theoretically, the desired stabilizing characteristic of reducing the influence of small $\{\lambda_k\}$. Convergence here can be very slow, however [3,4].

WELL-CONCENTRATED FUNCTIONS AND THEIR APPROXIMATE DIMENSION

We recall that the largest eigenvalue $\lambda_0 = \lambda_0(\Omega T)$ of (4) represents the maximum concentration

$$||D_T f||^2/||f||^2$$

attainable by a bandlimited function $f \in \underline{B}(\Omega)$. However, the
succeeding eigenvalues also have an interesting interpreta-
tion: by the minimax characterization [27, p.237],
$\lambda_{k-1} = \lambda_{k-1}(\Omega T)$ is the maximum k^{th} largest concentration
among k mutually orthogonal bandlimited functions. To pursue
the well-concentrated functions we therefore focus on the
behavior of eigenvalues; as these depend only on ΩT, let us
henceforth normalize so that $\Omega = \pi$. Since a projection never
increases the norm, we have $0 < \lambda_k < 1$, the strict inequalities a consequence of the analyticity of $f \in \underline{B}(\Omega)$. By applying to (4) the formulas for the trace and for the Hilbert-
Schmidt norm, we find

$$\sum_{k=0}^{\infty} \lambda_k(\pi T) = \frac{1}{\pi} \int_{-T}^{T} \pi \, dt = 2T, \tag{10}$$

$$\sum_{k=0}^{\infty} \lambda_k^2(\pi T) = \int_{-T}^{T}\int_{-T}^{T} \frac{1}{\pi^2} \frac{\sin^2 \pi(t-s)}{(t-s)^2} \, ds \, dt > 2T - c \log T, \tag{11}$$

for some c independent of T, so that

$$\sum_{k=0}^{\infty} \lambda_k(1-\lambda_k) < c \log T. \tag{12}$$

With $\delta > 0$ fixed let us now consider those $\{\lambda_k(\pi T)\}$ for which
$\delta < \lambda_k(\pi T) < 1 - \delta$. Each of these contributes an amount no
smaller than $\delta(1-\delta)$ to the left-hand side of (12), so that
the number of such contributions grows no faster than
$c_1 \log T$. Thus the eigenvalues are near 1, then near 0, the
transition occurring over a relatively narrow range of values
of k. Combining this with (10) and (11) shows that the
number of large eigenvalues is $2T - c_2 \log T$; the
corresponding eigenfunctions then span a subspace of $\underline{B}(\Omega)$ of
that dimension, in which every function is well-concentrated
on $|t| < T$. As for the intermediate eigenvalues, we can show
by a separate argument that, more precisely, $\lambda_{[2T]}(\pi T)$ is
bounded away from 0 and 1, independently of T [14], and
indeed – by suitable choice of the function h(t) figuring in
[14] – that

$$\lambda_{[2T]}(\pi T) \simeq 1/2.$$

(Here [x] denotes the integer part of x.)

On viewing $||f||^2$ as the energy of the function f(t), we can interpret the preceding results so as to suggest a possible resolution of the paradox of simultaneous time and band-limitedness of physical signals. For suppose that our reading of energy is accurate only to an order of magnitude ε, so that we cannot distinguish between 0 and a function h(t) for which $||h||^2$ is of the order of ε. This suggests, firstly, that a bandlimited function f(t) whose energy outside an interval $|t| < T$ lies below the detection threshold will appear time-limited. Moreover, any approximation whose accuracy is commensurate with this unconcentrated energy will likewise be perceived as exact. Specifically, for $f \varepsilon \underline{B}(\pi)$, if

$$f = \sum_{k=0}^{\infty} a_k \phi_k,$$

let

$$g = \sum_{k \leq [2T]-1} a_k \phi_k.$$

Then

$$||f-g||^2 = \sum_{k \geq [2T]} |a_k|^2 < \frac{1}{1-\lambda_{[2T]}} \sum_{k \geq 0} |a_k|^2 (1-\lambda_k)$$

$$< c \int_{|t|>T} |f(t)|^2 dt,$$

with c a fixed constant which, in view of the behavior of $\lambda_{[2T]}(\pi T)$, can be taken as 2. It follows that, whatever the precision with which energy can be measured, those band-limited functions which are perceived as time-limited (because the unconcentrated energy lies below the detection threshold) are likewise, to the same order of measurement accuracy, seen to lie in the [2T]-dimensional subspace spanned by $\phi_0, \ldots, \phi_{[2T]-1}$. It is this notion of approximate dimensionality - much better justified and explained in [31]-which expresses the engineering intuition mentioned at the outset.

It is interesting that there are simpler-generated [2T]-dimensional subspaces which approximate B(Ω) as well as does that spanned by the eigenfunctions. An example is the subspace spanned by the translates $\sin \pi(t-y_k)/(t-y_k)$, with

$\{y_k\}$ the zeros of $\phi_0(t)$ in $|t| < T$ [21]. This is a delicate matter, however, for the translates of sin $\pi t/t$ centered at the integers in $|t| < T$ produce a far poorer approximation [13].

APPROXIMATE PROJECTIONS; EIGENVALUE DISTRIBUTION

A self-adjoint operator whose eigenvalues are 1 and 0 is necessarily a projection. We have seen that this property is nearly true of $D_T B_\Omega D_T$, which we can therefore picture as being close to a projection onto the subspace of functions in $B(\Omega)$ well-concentrated on $|t| < T$, having dimension $2T\Omega/\pi - 0(\log T)$. This point of view helps to clarify results on the asymptotic distribution, as $T \to \infty$, of eigenvalues of the integral equation

$$\frac{1}{\sqrt{2\pi}} \int_{-T}^{T} k(x-y)f(y)dy = \mu f(x), \quad |x| < T ,$$

when k is integrable. For on subdividing the ω-axis into intervals I_j over which $K(\omega)$, the Fourier transform of $k(x)$, is nearly a constant, γ_j, we can approximate this integral operator by

$$\Sigma \ \gamma_j D_T B_{I_j} D_T.$$

This is a linear combination of nearly orthogonal approximate projections, so that the eigenvalues behave like γ_j with multiplicity approximately T meas $(I_j)/\pi$. Letting $T \to \infty$ shows that, to first order, these eigenvalues have the same distribution as do the values of $K(\omega)$ sampled at the integer multiples of π/T. This result was proved for real-valued $K(\omega)$ in [9], but the present argument applies also to complex-valued $K(\omega)$ [16,17].

A much more refined question concerns the behavior of the $\{\lambda_i\}$ of (4) in the transitional region $0 < \lambda < 1$. The conjecture of [30], that $\lambda[2T+(b/\pi)\log T] \to 1/(1+\exp(\pi b))$, was established in [18], and far-reaching generalizations are now known [1,36].

EXTENSIONS

      All but the very last of the results described depend
only on simple features of Hilbert space geometry, and so
apply without change when the intervals $|t| < T$ and $|\omega| < \Omega$,
to which time and frequency are limited, are replaced by
arbitrary compact sets, possibly in higher dimension [28, 29,
15, 6]. The structure also persists in discrete versions, as
for periodic functions, where Fourier coefficients play the
role of the Fourier transform [32]. Concentration and
approximation problems for band-limited functions can also be
posed in L^p spaces, $p \neq 2$, where – with orthogonal decomposition less ready to hand – they yield a rich variety of difficult questions. Extensive and deep results in this area have
been discovered by B.F. Logan, and will appear in a forthcom-
ing book [20]. Approximate dimension for bounded bandlimited
functions has been elegantly formulated and determined in
[22].

THE EIGENFUNCTIONS

      In view of their many potential applications, it is
valuable to find the eigenfunctions explicitly. The crucial
observation here is that the integral of (4) commutes with a
certain second-order differential operator, whence it follows
that $\{\phi_k\}$ are the prolate spheroidal wave eigenfunctions of
the latter [24]. A great deal of information, theoretical
and computational, about local and asymptotic behavior flows
from this identification; these results were derived with
dazzling virtuosity in [30]. A program for calculating the
$\{\phi_k\}$ is also available [23,35]. In discrete versions, where
band-limited sequences replace band-limited functions, a tri-
diagonal matrix plays the role of the commuting differential
operator, and permits rapid and accurate calculation of the
analogous successively best-concentrated eigensequences.

      Because of the light which the commutativity sheds on
the eigenfunctions, it is interesting to inquire when it can
take place. The question has been posed in generality by
F.A. Grunbaum [7,8] who, having discovered many examples in a
variety of analogous contexts, has asked for an explanation
that would account for them all. This problem reaches far,

and is still open.

SPECTRAL ESTIMATION

Spectral estimation seeks to produce a plausible decomposition into frequency components of a given function $f(t)$, generally obtained empirically over a finite interval: $f \in D(T)$ for some T. As the purpose is to elucidate the process which generated the data, interest often focuses on the extent to which this frequency content is concentrated on some subset of frequencies, not known in advance; in this way, the problem is connected with our subject. The best approximation to a function by a component having frequencies in an interval I is given by applying the projection B_I to the function, but as the sinx/x convolution kernel of B_I has considerable energy in its tail, this frequency-limiting operation is computationally inaccurate, especially for small I, and if used on $D_T f$ is too sensitive to the unobserved values of f in $|t| > T$. Convergence here can be improved by substituting for that kernel the best-concentrated eigenfunction ϕ_0 of $B_I D_U B_I$, with a suitable $U < T$, for since the energy of ϕ_0 in $|s| > U$ is small, the error contributed to the convolution of f and ϕ_0 by the time-limiting of f can be reduced in the range $|t| < T-U$; the choice of U effects a trade among the sizes of this range, of the error, and of I. The existence of an easily calculated accurate approximation to ϕ_0 makes this approach practical and useful [10, 11]. Recently, D.J. Thomson has suggested that the data could be utilized more fully, and in a way which has many desirable statistical characteristics, if the component of $f(t)$ with frequencies in I were approximated by projecting $D_T f$ onto the subspace spanned by the first few most-concentrated eigenfunctions of $B_I D_T B_I$. The total decomposition of f is then generated in this way by sweeping I over the frequencies. Here, when I_1 and I_2 are disjoint, the highly concentrated eigenfunction from $\underline{B}(I_2)$ are nearly orthogonal over $|t| < T$ to those from $\underline{B}(I_2)$, so that this extraction of frequency components proceeds by nearly orthogonal increments; when I_1 and I_2 overlap, an average is performed of the components determined in the common region. For the more usual case when the data consists of equally spaced samples, rather than of a continuous record, the eigenfunctions are replaced by eigensequences, which are readily calculated. This method has proved to be outstandingly successful [19, 34].

SAMPLING AND INTERPOLATION

The most important practical consequence of band-limitedness is that function of $\underline{B}(\Omega)$ are equivalent, in a stable way, to their values at the points $\{k\Omega/\pi\}$. Specifically, normalizing $\Omega = \pi$, if we write $F(\omega)$, $|\omega| < \pi$, as a Fourier series, we obtain

$$F(\omega) = \sum_{-\infty}^{\infty} a_n e^{-in\omega}/\sqrt{2\pi}, \tag{13}$$

where

$$a_k = \frac{1}{\sqrt{2\pi}} \int_{-\pi}^{\pi} F(\omega)e^{ik\omega}d\omega = f(k) , \tag{14}$$

$$\sum |f(k)|^2 = \sum |a_k|^2 = \int_{-\pi}^{\pi} |F(\omega)|^2 d\omega = \int |f(t)|^2 dt . \tag{15}$$

On applying the Fourier transform to (13) we find, by (14),

$$f(t) = \frac{1}{\sqrt{2\pi}} \int_{-\pi}^{\pi} F(\omega)e^{i\omega t}d\omega = \sum_{-\infty}^{\infty} f(n) \frac{\sin \pi(t-n)}{\pi(t-n)} , \tag{16}$$

the series converging in L^2, and, by Schwarz's inequality, also uniformly. The formula (16) is thus an expansion in the orthonormal basis of functions $\{\sin \pi(t-n)/\pi(t-n)\}$, which simultaneously form an interpolating family at the sampling points $\{t = n\}$. The orthonormality ensures stability, for an imprecision ε_n in reading the sample value $f(n)$ will lead to a reconstruction whose error,

$$e(t) = \sum \varepsilon_n \sin \pi(t-n)/\pi(t-n),$$

will, by (15), have norm equal to the measurement error $\sum |\varepsilon_n|^2$. The interpolating property of the basis functions ensures that arbitrary square-summable values can likewise be stably realized as the sample values of a simply constructed band-limited function of finite energy. Thus (16) provides a stable way of passing between continuous band-limited and discrete information - the foundation on which modern telecommunication systems are built. The rate of sampling, Ω/π, at which this is done is often called the Nyquist rate.

In view of the usefulness of (16), it is natural to ask whether functions of $\underline{B}(\Omega)$ can be recovered from their values at other, perhaps sparser, sets of points; this would make discretization of band-limited signals more efficient. Likewise, we can try to find denser sets of points at which arbitrary square-summable values $\{a_k\}$ can be interpolated by some $f \in \underline{B}(\Omega)$; this would make band-limited transmission of discrete information more efficient.

To examine these questions, we consider the following possible features of a given set of real points Λ.

A. Λ is a set of uniqueness for $\underline{B}(\Omega)$ if each $f \in \underline{B}(\Omega)$ must vanish identically if it vanishes at all the points of Λ.

B. Λ is a set of stable sampling for $\underline{B}(\Omega)$ if the points of Λ are uniformly separated (i.e., there is $\delta > 0$ such that for distinct $\tau_i, \tau_j \in \Lambda, |\tau_i - \tau_j| > \delta$) and if there is a constant K such that, for each $f \in \underline{B}(\Omega)$,

$$\int_{-\infty}^{\infty} |f(t)|^2 dt < K \sum_{\tau \in \Lambda} |f(\tau)|^2.$$

C. $\Lambda = \{\tau_k\}$ is a set of interpolation for $\underline{B}(\Omega)$ if, corresponding to each square-summable sequence $\{a_k\}$, there exists $f \in \underline{B}(\Omega)$ with $f(\tau_k) = a_k$.

If Λ is a set of uniqueness for $\underline{B}(\Omega)$, then two functions of $\underline{B}(\Omega)$ which agree at the points of Λ must coincide; thus the sample values $\{f(\tau), \tau \in \Lambda\}$, determine f everywhere. Sets of uniqueness can be very sparse. For example, invoking the analyticity of f, we see that any collection of points with a point of accumulation will qualify. To exhibit more uniformly distributed Λ, we can take integer multiples of $\alpha > \Omega/\pi$, only on a half-line. There are many kinds of examples, and their variety seemed chaotic until, in what is undoubtedly the deepest result in a vast area, A. Beurling and P. Malliavin discovered a density whose size controls the feature of uniqueness [2,26].

Although f ∈ $\underline{B}(\Omega)$ is determinable from its values on a
set of uniqueness, this fact does not constitute a sampling
reconstruction adequate in practice. For, just as in the
problem of extrapolating from a finite interval, an arbitrarily small error in measuring the sample values can well
lead to arbitrarily large errors in the reconstructed signal,
or indeed can make the extrapolation impossible. In order to
ensure stability of reconstruction, we require that Λ be a
set of stable sampling for $\underline{B}(\Omega)$. In sharp contrast to sets
of uniqueness, the points of a set of stable sampling must be
regularly distributed and, on all sufficiently large intervals, must have at least the Nyquist rate [14]. For an
intuitive explanation, let us take an interval I of length r,
let n(r) be the number of $\{\tau_k\}$ in I, and let us consider the
set S of function of $\underline{B}(\Omega)$ well-concentrated on I, so that the
energy of f ∈ S outside I is small. If we can presume that
the samples of f ∈ S at the points $\{\tau_k\}$ outside I are also
small, then, by virtue of stability, replacing these sample
values by zero should not greatly affect our reconstruction
of f(t). We conclude that a function of S is substantially
determined by n(r) measurements, so that the collection S is,
in a sense, no more than n(r)-dimensional. But we have shown
the approximate dimension of S to be at least
r Ω/π - c log r. Thus n(r) ⩾ r Ω/π - c log r from which
$\lim_{r \to \infty} n(r)/r$ ⩾ Ω/π, and our assertion follows.

For sets of interpolation, similar considerations apply,
and show that the maximum density on all large intervals cannot exceed Ω/π. Thus the Nyquist rate for sampling and
interpolation of band-limited functions cannot be improved
[14]. Again, since the argument is based only on the first-
order behavior of the eigenvalues of (4), it can be carried
out without change for more complicated frequency sets than a
single interval, and in more dimensions [15].

CONCLUSION

Band-limited functions possess many properties that stem
from their analyticity. However, as analyticity is fragile,
not all of these persist under small perturbation. If we
require that our conclusions remain stable when functions are
determinable only with given precision, we are led to problems in which the time-and-frequency-limiting operator enters

naturally. The eigenvalues of this operator show that, to
first order, it resembles a projection, and the stable properties of band-limited functions typically inherit from it a
simple and orderly behavior, useful in theory and application.

REFERENCES

1. E. Basor and H. Widom, Toeplitz and Wiener-Hopf deter-
    minants with piecewise continuous symbols, J. Functional
    Analysis 50 (1983), 387-413.

2. A. Beurling and P. Malliavin, On the closure of charac-
    ters and zeros of entire functions, Acta Math. (Stock-
    holm) 118 (1967), 79-93.

3. P. DeSantis and F. Gori, On an iterative method for
    super-resolution, Optica Acta 22 (1975), 691-695.

4. P. DeSantis, F. Gori, G. Guattari, and C. Palma, Optical
    systems with feedback, Optica Acta 23 (1976), 505-518.

5. R.W. Gerchberg, Super-resolution through error energy
    reduction, Optica Acta 21 (1974), 709-720.

6. A. Papoulis, A new algorithm in spectral analysis and
    bandlimited extrapolation, IEEE Trans. Circuits Syst.,
    CAS-22 (1975), 735-742.

7. F. Gori and S. Paolucci, Degrees of freedom of an opti-
    cal image in coherent illumination, in the presence of
    observations, J. Opt. Soc. Am 65 (1975), 495-501.

8. F.A. Grunbaum, Finite convolution integral operators
    commuting with differential operators: some counterex-
    amples, Num. Funct. Anal. and Optimiz. 3 (1981), 185-
    199.

9. F.A. Grunbaum, Band and time limiting, recursion rela-
    tions and some nonlinear evolution equations, in "Spe-
    cial Functions: Group Theoretical Aspects and Applica-
    tions," R. Askey, T. Koornwinder and W. Schemp, eds,
    Reidel, Dordrecht (1984).

10. M. Kac, W. Murdock, and G. Szego, On the eigenvalues of
     certain Hermitian forms, J. Rat. Mech. Analysis 2
     (1953), 767-800.

11. J.F. Kaiser, Nonrecursive digital filter design using
     the I_0-sinh window function, Proc. 1974 IEEE Symp. Cir-
     cuits and Systems, 20-23.

12. J.F. Kaiser and R.W. Schafer, On the use of the I_0-sinh
     window for spectrum analysis, IEEE Trans. Acoust.,
     Speech and Signal Processing, ASSP-28 (1980), 105-107.

13. H.J. Landau and H.O. Pollak, Prolate spheroidal wave
     functions, Fourier analysis and uncertainty II, Bell
     Syst. Tech. J. 40 (1961), 65-84.

14. H.J. Landau and H.O. Pollak, Prolate spheroidal wave
     functions, Fourier analysis and uncertainty III, Bell
     Syst. Tech. J. 41 (1962), 1295-1336.

15. H.J. Landau, Sampling, data transmission, and the
     Nyquist rate, Proc. IEEE 55 (1967), 1701-1706.

16. H.J. Landau, Necessary density condition for sampling
     and interpolation of certain entire functions, Acta
     Math. (Stockholm) 117 (1967), 73-93.

17. H.J. Landau, On Szego's eigenvalue distribution theorem
     and non-Hermitian kernels, J. d'Analyse Math. 28
     (1975), 335-357.

18. H.J. Landau, The notion of approximate eigenvalues
     applied to an integral equation of laser theory, Quart.
     Appl. Math. 35 (1977), 165-172.

19. H.J. Landau and H. Widom, The eigenvalue distribution of
     time and frequency limiting, J. Math. Anal. and Appl.
     77 (1980), 469-481.

20. C. Lindberg, J. Park, and D.J. Thomson, Optimal spectral
     windows for normal mode time series, American Geophysi-
     cal Union, Fall meeting, San Francisco, Dec. 1983.

21. B.F. Logan, "High-Pass Functions", to appear.

22. A.A. Melkman, N-widths and optimal interpolation of
     time- and band-limited functions, "Optimal Estimation in
     Approximation Theory", C.A. Micchelli and T.J. Rivlin,
     eds, Plenum, New York (1977), 55-68.

23. A.A. Melkman, N-widths and optimal interpolation of
     time- and band-limited functions II, to appear.

24. B.J. Patz and D.L. Van Buren, A FORTRAN computer program
     for calculating the prolate spheroidal angular functions
     of the first kind, Naval Research Laboratory Report 4414
     (1981), Washington, D.C.

25. H.O. Pollak and D. Slepian, Prolate spheroidal wave
     functions, Fourier analysis and uncertainty I, Bell
     Syst. Tech. J. 40 (1961) 43-64.

26. J.F. Price, Inequalities and local uncertainty princi-
     ples, J. Math. Phys. 24 (1983), 1711-1714; Sharp local
     uncertainty inequalities (submitted).

27. R.M. Redheffer, Completeness of the set of complex
     exponentials, Advances in Math. 24 (1977), 1-62.

28. F. Riesz and B. Sz-Nagy, "Functional Analysis," Ungar,
     New York (1971).

29. D. Slepian, Prolate spheroidal wave functions, Fourier
     analysis and uncertainty IV, Bell Syst. Tech. J. 43
     (1964), 3009-3057.

30. D. Slepian, Analytic solution of two apodization prob-
     lems, J. Opt.Soc. Am. 55 (1965), 110-115.

31. D. Slepian, Some asymptotic expansions for prolate
     spheroidal functions, J. Math. and Phys. 44 (1965), 99-
     140.

32. D. Slepian, On bandwidth, Proc. IEEE 63 (1976), 292-300.

33. D. Slepian, Prolate spheroidal wave functions, Fourier
     analysis and uncertainty V, Bell Syst. Tech. J. 57
     (1978), 1371-1430.

34. D. Slepian, Some comments on Fourier analysis, uncer-
     tainty and modeling, SIAM Review 25 (1983), 379-393.

35. D.J. Thomson, Spectrum estimation and harmonic analysis,
    Proc. IEEE 70 (1982), 1055-1096.

36. A.L. Van Buren, A FORTRAN computer program for calculat-
    ing the linear prolate functions, Naval Research Labora-
    tory Report 7994 (1976), Washington, D.C.

37. H. Widom, On a class of integral operators with discon-
    tinuous symbol, in : "Toeplitz Centennial (Tel-Aviv,
    1981), Operator Theory: Adv. Appl., 4," I. Gohberg, ed.,
    Birhauser, Basel-Boston, (1982), 477-500.

FROM RAINBOWS TO RINGS: A HISTORY OF THE IDEA OF THE SPECTRUM

D.E. Taylor

Department of Pure Mathematics
University of Sydney
Sydney, N.S.W.
Australia

OUTLINE

In twentieth century mathematics the word ´spectrum´ occurs in many places and is used to describe a wide variety of apparently unrelated ideas. The evolution of the meaning of this word begins with the rainbow of light created by a prism and passes by way of integral equations and quantum mechanics to the space of prime ideals of a ring. Along the way one glimpses much of modern harmonic analysis and operator theory.

The spectrum of colours produced by a prism or diffraction grating can be regarded as the Fourier transform of the incoming waveform. In the 1860´s Maxwell showed that light was but a small part of a vast electromagnetic spectrum. The discovery of radio waves by Hertz and the use of X-ray diffraction in crystallography confirmed this and so the notion of spectrum was extended to this wider domain. Eventually the word spectrum was applied to the plot of amplitude against frequency obtained by taking the Fourier transform of any function for which it was defined.

An atom, when suitably excited, emits radiation whose spectrum consists of a number of ´sharp lines´ characteristic of the atom in question. In 1917 Bohr found that this phenomenon could be explained in terms of the quantum ideas introduced by Planck. In 1925 Heisenberg and Schrödinger independently incorporated Bohr´s rules into a coherent

quantum theory. Four years later von Neumann showed these
theories to be equivalent. In von Neumann's formulation each
physical observable is represented by a self-adjoint operator
on a Hilbert space and it is the set of eigenvalues of the
energy operator which determines Bohr's 'stationary states'
of the atom; the frequency spectrum of the atom is obtained
by taking differences of eigenvalues. The spectrum of an
operator T consists of those complex numbers λ such that
$T - \lambda I$ does not possess an inverse. The word 'spectrum' was
already in use in this context in a 1926 paper on quantum
mechanics by Born, Heisenberg and Jordan. Chapter 3 of this
paper was written by Born, who was familiar with the work of
Hilbert on integral equations and quadratic forms in infinitely many variables. This allowed him to treat line spectra
and continuous spectra on an equal footing. Hilbert had
defined the point spectrum and the continuous spectrum of a
quadratic form in 1906 but later expressed surprise that his
mathematical spectrum coincided with the physical spectrum.

A major result in Hilbert's work on integral equations
is his theorem expressing a completely continuous quadratic
form as an integral over the spectrum. This generalized the
'Principal Axes Theorem' on the reduction of a finite
quadratic form to a sum of squares and developed into the
'Spectral Theorem' of von Neumann, Stone and others. One
introduces a projection valued measure Π (which depends on
the operator T), defined on the Borel sets of the spectrum
and then expresses T as an integral $\int \lambda d\Pi$. The measure Π is
an isomorphism between the Boolean lattice of Borel sets of
the spectrum and a lattice of orthogonal projections on the
Hilbert space. In this way the notion of spectrum can be
transferred to measures (simply meaning the support of the
measure) and, more significantly, to Boolean lattices.
Indeed this led Stone (in 1935) to consider Boolean lattices
in general and to study their representations as lattices of
subsets of a set. Stone observed that Boolean lattices are
equivalent to Boolean rings (commutative rings in which every
element is idempotent). The connection between the ring and
lattice operations is given by $xy = x \wedge y$ and $x + y = (x \wedge y') \vee (x' \wedge y)$. In the case of the measure Π, the points of
the spectrum correspond to the maximal ideals of the Boolean
ring; hence the name 'spectrum' for the set of maximal ideals
of a general Boolean ring. Stone obtained his representation
theorem for Boolean lattices by first passing to the Boolean
ring and then associating each element to the set of maximal
ideals which contain it. In another paper Stone introduced a

topology on the spectrum by taking for each ideal the set of
maximal ideals which contain it and calling these the closed
sets.

These ideas were soon extended to the theory of commutative Banach algebras by Gelfand and his school. The ring of
all continuous complex valued functions defined on a compact
Hausdorff space X is a commutative Banach algebra C(X) and
the set of maximal ideals of C(X), topologized as above, is
homeomorphic to the original space X. Gelfand showed that,
for any commutative Banach algebra B, this topology makes the
set Max(B) of maximal ideals into a compact Hausdorff space
and there is a natural homomorphism from B into C(Max(B)).
In general this map is neither one-to-one nor onto.

If T is a bounded self-adjoint operator on a Hilbert
space, then T generates a commutative Banach algebra B. The
space X = Max(B) is homeomorphic to the spectrum of T and B
is isometrically *-isomorphic to C(X). Thus X can be considered as a compact subspace of the real line and the element of C(X) corresponding to T is simply the function
f(x) = x. All these considerations extend to the case of a
normal operator.

In ring theory Jacobson showed that one could extend
Stone's definition of a topology to arbitrary rings provided
one took the set of all ´primitive´ two-sided ideals instead
of the set of all maximal ideals. (If the ring is commutative the primitive ideals coincide with the maximal ideals.)
At about the same time Zariski defined a topology on the set
of all places of a function field by analogous method. Weil
showed that all algebraic varieties can be given a topology
of this type and that this allows one to introduce the notion
of fibre space and vector bundle in analogy with differentiable manifolds. Over an algebraically closed field the
points of the variety correspond to the maximal ideals of the
coordinate ring. Serre used these ideas to introduce the
theory of sheaves and cohomology to the study of algebraic
varieties.

Around 1957 it was realized by Grothendieck and others
that the passage from a commutative ring to its space of maximal ideals does not define a functor and that instead of
maximal ideals one should study the prime ideals. The topology is defined as before. To each ideal of the ring one
associates the set of all prime ideals which contain it:

these are the closed sets. This is now called the Zariski
topology and the set of all prime ideals of the ring A fur-
nished with this topology is called the spectrum, Spec(A), of
A.

One can go further and associate to each (basic) open
set of Spec(A) a suitable localization of A in such a way
that the resulting structure is a sheaf of rings. Ringed
spaces of this form may be glued together to form ´schemes´.
These are the objects of study in modern abstract algebraic
geometry.

Commutative algebra generalizes both algebraic geometry
and algebraic number theory and the theory of schemes is a
further generalization. The study of schemes and sheaves led
Grothendieck to the related notions of site (a category with
a ´topology´) and topos (a special type of category general-
izing a sheaf). It turns out that topoi represent a fusion
of both geometry and logic and that the notion of spectrum
can be generalized to this context. This has been carried
out by Hakim, Lawvere, Joyal, Tierney and others. It can be
thought of as the construction of a ´free local ring´ on a
given ring object of the topos. In carrying out the construction one is forced to change to a new topos.

The development from the rainbow to topos theory covers
most of the uses of ´spectrum´ within mathematics. There are
of course other occurrences such as the ´spectral sequences´
and ´spectra of spaces´ of algebraic topology. These uses
tend to conform to the common meaning of ´a range of interre-
lated values, objects etc.´ and are not connected with eigenvalues of operators.

REFERENCES

1. J. Dieudonné, History of Functional Analysis, Notas de
    Matemática (77), North-Holland (Amsterdam, New York,
    Oxford, 1981).

2. L.A. Steen, "Highlights in the history of spectral
    theory", Amer. Math. Monthly 80 (1973) 359-381.

LIST OF CONTRIBUTORS

R.H.T. Bates, Electrical and Electronic Engineering Department, University of Canterbury, Christchurch, New Zealand

J.J. Benedetto, Department of Mathematics, University of Maryland, College Park, Maryland 20742, USA

R.N. Bracewell, Department of Electrical Engineering, Stanford University, Stanford, California 94305, USA

G. Brown, School of Mathematics, University of New South Wales, Kensington, NSW 2033, Australia

P.W. Buchen, Department of Applied Mathematics, Sydney University, Sydney, NSW 2006, Australia

T.W. Cole, Department of Electrical Engineering, Sydney University, Sydney, NSW 2006, Australia

H.J. Landau, A.T. & T. Bell Laboratories, Murray Hill, New Jersey 07974, USA

J.F. Price, School of Mathematics, University of New South Wales, Kensington, NSW 2033, Australia

J.W. Sanders, School of Computing Sciences, New South Wales Institute of Technology, Broadway 2007, Australia

T.P. Speed, Division of Mathematics and Statistics, Commonwealth Industrial and Scientific Research Organization, Yarralumla, ACT 2600, Australia

D.E. Taylor, Department of Pure Mathematics, Sydney University, Sydney, NSW 2006, Australia

INDEX